POLYMER SCIENCE AND TECHNOLOGY

POLYANILINE

FROM TRADITION TO INNOVATION

POLYMER SCIENCE AND TECHNOLOGY

Additional books in this series can be found on Nova's website
under the Series tab.

Additional e-books in this series can be found on Nova's website
under the e-book tab.

POLYMER SCIENCE AND TECHNOLOGY

POLYANILINE

FROM TRADITION TO INNOVATION

CRISTINA DELLA PINA
AND
ERMELINDA FALLETTA

New York

NOTICE TO THE READER

LIBRARY OF CONGRESS CATALOGING-IN-PUBLICATION DATA

ISBN: 978-1-63463-273-7

Published by Nova Science Publishers, Inc. † New York

This Book is dedicated to our Mentor,
Professor Michele Rossi

Extraordinary Master of Science and Life
Forever grateful
Cristina & Ermelinda

Contents

ACKNOWLEDGMENTS

We wish to heartily thank Nova Science Publishers Inc. and, particularly, its President Mrs. Nadya Gotsiridze-Columbus for kindly inviting us to write this book.

Cristina & Ermelinda

INTRODUCTION: POLYANILINE: THE IMPORTANCE OF STRUCTURE, CRISTALLINITY DEGREE, MOISTURE, MOLECULAR WEIGHT AND *DOPING* FOR TUNING CONDUCTIVITY

Among the intrinsically conducting polymers (ICPs), polyaniline (PANI) is one of the oldest and most investigated. Even though the date of the first synthesis is not sure, in 1834 Runge described a method to isolate from coal tar a substance that became blue colored when treated with chloride of lime and he named it kyanol or cyanol. [1] In 1856 by the oxidation of crude aniline in the presence of potassium dichromate the first synthetic dye, mauveine, was industrially obtained. [2-4] A black precipitate, known as "aniline black", was also observed in 1860. [5] It was a polymeric constituent of melanin. However, only the accurate investigations of Green and Woodhead allowed to better understand the properties of this surprising material. [6,7]

Over the years the interest in polyaniline has rejuvenated owing to its peculiar chemico-physical characteristics, such as unique doping/dedoping mechanism, environmental and thermal stability, and tuneable conductivity. This book will focus on polyaniline. After an introduction on its various structures available, with an emphasis on the importance of the different oxidation states, *doping* (protonation), degree of crystallinity, moisture and molecular weight for tuning conductivity, the main synthetic methods will be presented. They will range from the traditional protocols to the innovative eco-friendly routes which employ "green" oxidants and proper catalysts. Regarding the ultimate preparation methods, a particular focus will be dedicated to the starting reagents. It has been recently found how starting from the monomer (aniline) or the dimer (AD, *N*-(4-aminophenyl)aniline) deeply affects polymerization and properties of the final material, with important repercussions on the potential applications. The choice of the catalyst is also fundamental, thus another section will cover the state-of-the-art of the novel catalytic systems. The doping agents, degree of crystallinity and moisture, whose effects on PANI conductive capability are dramatic, deserve particular attention therefore recent findings on their influence over the final material performance will be mentioned as well. The possibility to combine the fascinating conducting properties of PANI with those of other materials has opened the way to the preparation of new amazing materials. In this context, organic–inorganic nanocomposites with organized structure have been extensively studied, because they combine the advantages of inorganic materials (mechanical strength, electrical and magnetic properties and thermal stability) with those of organic polymers (flexibility, dielectric behavior, ductility and processing). Accordingly, a chapter will describe the principal synthetic methods and properties of PANI-based composites. For completing the part dedicated to polyaniline blends, a special note will describe a brand new preparation protocol able to facilitate polymer processing into pure electrospun nanofibers. The main analytical techniques suitable for characterizing polyaniline and its composites will be collected in a further chapter. The spectrum of applications of PANI-based materials is wide, ranging from electrical and optical devices to sensors, as well as EMI shielding, anticorrosive and biomedical materials. A short description of two important accomplishments in this area will find place in a dedicated chapter, followed by the conclusion. Academics, researchers, scientists, engineers and students dealing with material science and nanotechnology are the target audience for this book.

THE IMPORTANCE OF STRUCTURE

Polyanilines refer to a class of polymers composed of repeated aniline units connected to each other to form a backbone, whose basic form has the general structure reported in Figure 1.

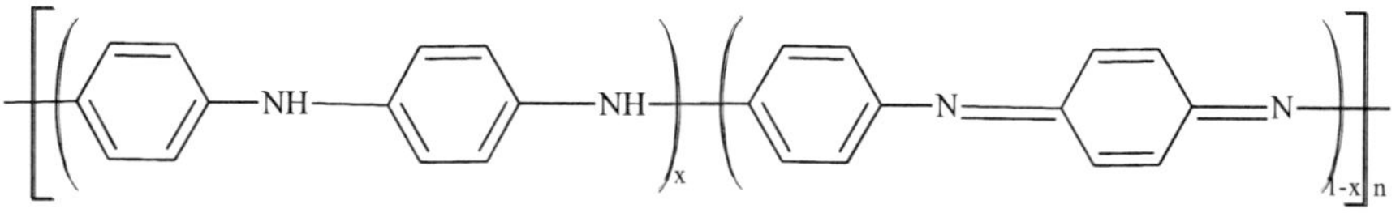

Figure 1. Polyaniline general structure.

The oxidation state (x) can vary from 1, giving the totally reduced form called *leucoemeraldine*, to 0.5 thus producing the half-oxidized form, *emeraldine*, as well as zero which leads to the completely oxidized form, *pernigraniline*. Besides these three main structures, further intermediate forms are available, such as *protoemeraldine*, characterized by an oxidation state intermediate between *leucoemeraldine* and *emeraldine*, and *nigraniline*, having an oxidation degree between *emeraldine* and *pernigraniline.* [8]

Each form can exist either as base or as protonated salts. However, only the protonated salt form of *emeraldine* (*emeraldine salt*, ES) is conductive. In fact, the complete protonation of the nitrogen atoms of imine groups leads to the formation of delocalized polysemi-quinone radical cations that increase the electronic conductivity of the polymer. Other chemico-physical properties of polyaniline are strictly correlated to its degree of oxidation, for example optical, electrical and thermal aspects.

More in detail, *leucoemeraldine* is an amorphous material with a white/pale brown colour, not stable under air and, if heated, it undergoes quick oxidization to *protoemeraldine.*

The colour of *protoemeraldine* ranges from violet in its base form to yellowish/pale green if protonated.

In its base form *emeraldine* is blue coloured and soluble in some organic solvents, as pyridine, *N*, *N*-dimethylformamide and *N*-methylpyrrolidinone. It becomes green after doping (protonation) with an acid. Undoped *nigraniline* is characterized by a dark blue coloration. This form is unstable at high temperature passing to the more stable form of *emeraldine.*

In its highest oxidation state, *pernigraniline*, polyaniline is not stable quickly decomposing to form lower oxidation state species.

However, among all the possible PANI structures, only the half-oxidized *emeraldine* exhibits conducting properties when it is protonated with inorganic or organic acids.

A GLANCE TO THE PARAMETERS INFLUENCING THE CONDUCTIVITY

The most intriguing property of polyaniline is being able to switch from an insulating state to a conducting one, thereby opening the way for many applications especially in electronics. In the last two decades scientists all over the world have suggested possible theoretical explanations for PANI conduction mechanism, proposing the heterogeneous disorder theory, [9] the phonon theory, [10] the band theory, hole and hopping theory, just for citing the principal ones. However, the ability of PANI to conduct electricity is the sum of many factors and, therefore, several parameters have been monitored, such as the oxidation degree of polymeric chains, the doping level, the crystallinity degree, the humidity level and the molecular weight. More emphasis on all these aspects will be given in Chapter 1. In principle, after protonation, the partially oxidized form of *emeraldine* undergoes an electron rearrangement thus producing stable radical cations, called polarons, that are the real active centres in the conducting process. Figure 2 displays the rearrangement of electrons in the backbone of *emeraldine* salt.

1)

Polarons formation

2)

Polarons delocalization

3)

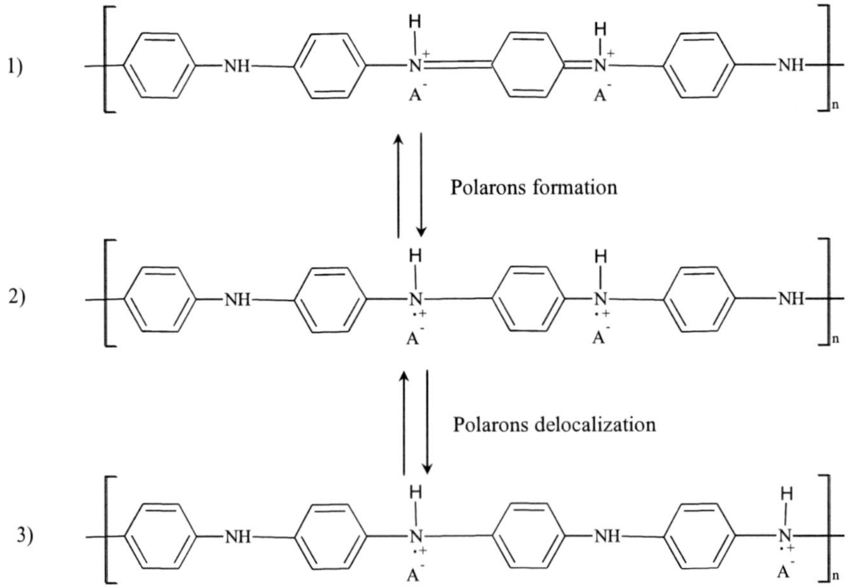

Figure 2. Electrons rearrangement in the backbone of *emeraldine* salt.

In step 1 an electron is transferred from each imine nitrogen into the aromatic ring leading to a benzenoid system. On each nitrogen atom this process leaves a single unpaired electron, that represents a radical cation, called polaron, that is unstable, and for this reason evolves towards the formation of two polarons (Figure 2, steps 2 and 3). The polaron lattice is responsible for the conductivity of polyaniline. [11]

Differently from other conducting polymers, characterized by a n-type and a p-type doping, the main doping process of polyaniline consists of a polymer protonation with Brønsted acids (protonic acid doping). In this context, the correlation between pH of the solution and increase of electrical conductivity in *emeraldine* salt, as well as the influence of the type of dopant on conductivity values were pointed out. [12] The crystallinity degree dramatically affects the electrical properties of polyaniline as conductivity increases with crystallinity, owing to a more organized structure of the polymeric chains. [13]

The correlation between conducting properties and humidity level was investigated by many researchers that suggested different models to explain the positive effect of water molecules on PANI conductivity. [14, 15]

Regarding the molecular weight, it has been found that too long chains can entail a number of structural imperfections, such as distortions and branching, thus depressing the polymeric conductivity. [16]

Chapter 1

SYNTHESIS OF POLYANILINE: FROM THE TRADITIONAL TO THE INNOVATIVE "GREEN" PROTOCOLS

Over the years a huge number of methods have been proposed to synthesize polyaniline and its derivatives. However, all these approaches can be summarized in two main categories: electrochemical and chemical routes. Owing to the environmental constraints, invoking to avoid polluting reagents, and the growing demand for highly pure materials, innovative catalytic approaches are offering "green" ways towards polyaniline.

In this chapter both the main traditional and the innovative methods for polyaniline preparation will be discussed with a particular focus on the chemical approach.

1.1. STANDARD PREPARATION METHOD: THE ELECTROCHEMICAL SYNTHESIS

If the chemical polymerization results to be the best method when a large quantity of polyaniline is requested, on the other hand the electrochemical polymerization is useful to produce the polymer in small amount but with a high purity and control of morphology and conductivity. This is especially required in the preparation of thin conducting polymeric films, particularly indicated to prepare electrodes for sensing applications.

It was demonstrated that the experimental conditions, such as electrode material, kind of dopant, pH value, solvent and so on strongly affect the polymerization process. [17] In particular, to achieve polyaniline in its conductive *emeraldine* form low pH values are requested. On the contrary, when the reaction is carried out at higher pH values, the deposited film contains only no conducting oligomers. [18] Moreover, the value of conductivity, morphology and rate of polyaniline growth are strongly influenced by the kind of doping agent used. [19]

Similarly to the oxidative approach, also the electrochemical polymerization of aniline is carried out only under acidic conditions, since high pH values lead to no conducting oligomers.

In general, the mechanism of any polymer synthesis - polyaniline included - involves three main steps: (1) initiation, (2) chain propagation and (3) termination.

(1) Initiation. The first step of aniline polymerization implies the oxidation at the electrode surface which produces aniline radical cations, as described in Figure 1.1. [20]

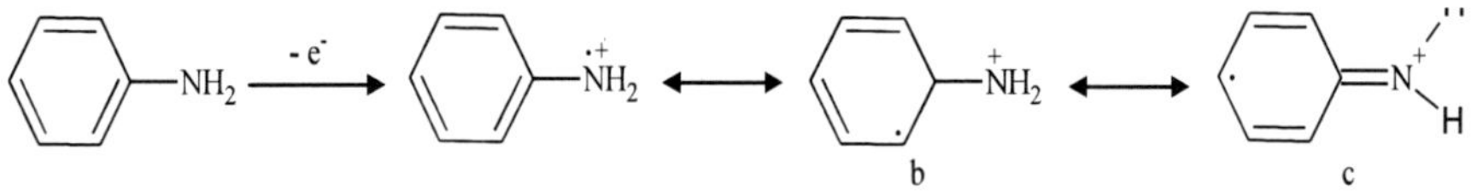

Figure 1.1. Aniline radical cations formation.

The formed aniline radical cation has several resonant forms, where (c) is the more reactive form due to its important substituent inductive effect and absence of steric hindrance.

The existence of aniline radical cations and, as a result, a radical mechanism was abundantly confirmed by adding into the reaction mixture radical inhibitors able to stop or retard the reaction. [21] These radical species can undergo different reactions depending on their reactivity and experimental conditions (i.e., temperature, electrolyte composition, deposition current density, potential scan rate, kind of anode material). Accordingly, they can interact with anions or solvent molecules to achieve products characterized by high solubility and low molecular weight. [22] Alternatively, they can undergo a coupling reaction involving the elimination of two protons and the re-aromatization of the rings, thus producing dimeric species whose nature is crucial both for the polymerization progress and the material properties (Figure 1.2).

Benzidine and *p*-aminodiphenylamine (PADPA) are the products of the tail-to-tail and head-to-tail coupling reactions respectively. [23] Both of them can initiate the polymerization reaction and be included in the polymeric structure. However, owing to its carcinogenic properties, the production of benzidine represents a concern for practical applications of the polymer. [24]

(A) $\dot{N}H_2^{+}$ → H_2N–C₆H₄–C₆H₄–NH_2

Benzidine

(B) $\dot{N}H_2^{+}$ → NH, H_2N

p-aminodiphenylamine

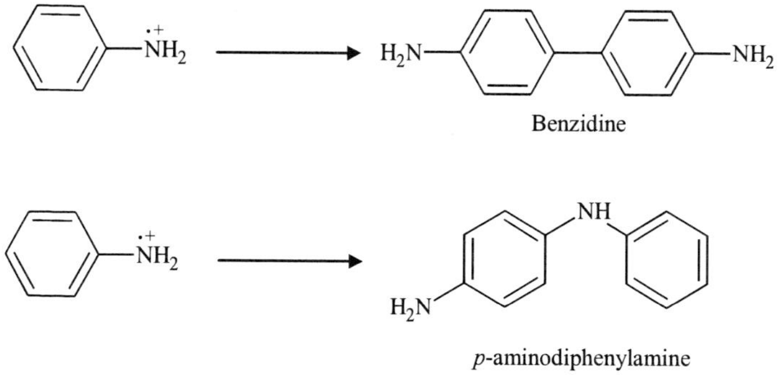

Figure 1.2. Coupling reaction producing (A) benzidine and (B) 4-aminodiphenylamine.

The existence of the intermediate PADPA was confirmed by FT-IR spectroscopy. Subsequent investigations of Mohilner et al. showed that the electrochemical oxidation of PADPA is easier than aniline as such, because it generally occurs during the dimerization reactions. [25] For this reason many authors concluded that the aniline oxidation step governs both efficiency and kinetics of the polymerization reaction.

(2) Chain propagation. Similarly to aniline monomer, aniline dimer, as well as longer species (oligomers), can be oxidized by anodic oxidation on the electrode surface leading to the corresponding radical cations able to react with aniline monomer. (Figure 1.3)

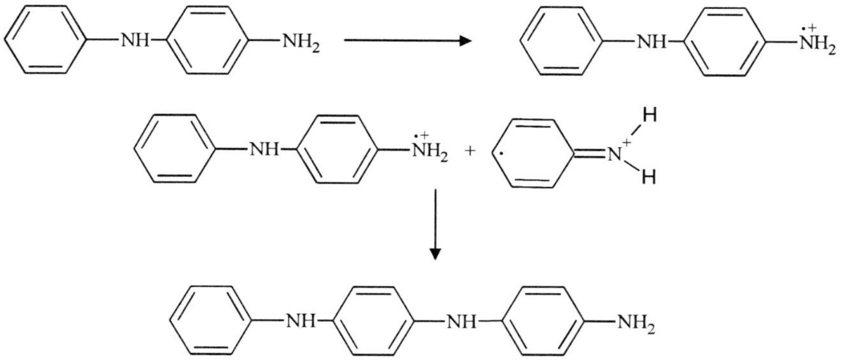

NH–NH_2 → NH–$\dot{N}H_2^{+}$

NH–$\dot{N}H_2^{+}$ + N^{+} H H

NH–NH–NH_2

Figure 1.3. Polyaniline chains propagation.

Although some authors suggested a similar polymerization mechanism for aniline and PADPA, [26] subsequent results highlighted that the electropolymerization of aniline dimer (PADPA) leads to polymers having lower molecular weights than those obtained by the monomer oxidation. [27]

The polymer chain obtained *via* the above-described coupling is formally in the most reduced state (*leucoemeraldine* form). However, during the propagation step, it undergoes further oxidation leading to polymeric chains with the characteristic form of *pernigraniline*, as schematically depicted in Figure 1.4.

The growth of polyaniline chains is guaranteed only at pH values lower than 2.5, otherwise polymerization does not occur. The polymeric chains, initially obtained in the form of unstable protonated *pernigraniline* characterized by a dark blue coloration, quickly turn to the more stable *emeraldine* form as displayed in Figure 1.5. [18]

The mechanisms described above (Figures 1.4 and 1.5) are simplifications of the more complex reactions that may occur.

In fact, different mechanisms might contribute to the polymer growth depending on the reaction conditions. More in particular, the polymerization mechanism, ruled by the applied potential of electropolymerization, is related to the presence of aniline monomers in form of neutral molecules, radical cations or cations and can be summarized as shown in Figure 1.6.

The mechanism reported in Figure 1.6 can be distinguished in: (i and ii) nucleophilic addition of the aniline monomer, (iii) radical addition and (iv and v) electrophilic addition.

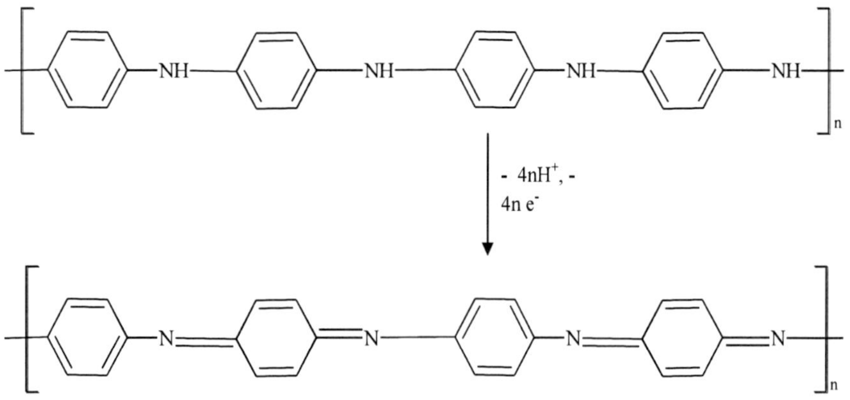

Figure 1.4. Oxidation of *leucoemeraldine* to *pernigraniline*.

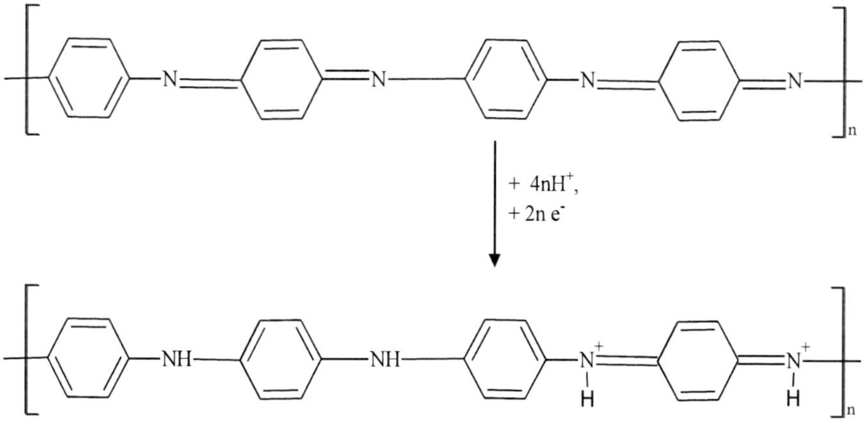

Figure 1.5. Reduction and protonation from *pernigraniline* to the more stable *emeraldine.*

Steps i and ii have lower reactivity than steps iii-v, because the charge density is more delocalized in the polymer than in the monomer. For this reason, at the beginning of aniline electropolymerization, steps iii-v are dominant. However, the ohmic drop increases with the thickness of the film, thereby producing a decrease in the applied potential at the solution-polymer interface and consequently favoring steps i and ii. [24]

(3) Termination. The final step in the polymerization reaction is not clear and many hypotheses have been formulated. It is argued that the polymerization reaction stops by hydrolysis of the terminal amino group, but the growth of chains can be stopped also because radical cations gradually become relatively unreactive owing to steric hindrance. The discussion is still open.

1.2. Standard Preparation Method: The Chemical Synthesis

While many efforts have been addressed to investigate the mechanism of aniline electrochemical polymerization and the parameters affecting it, much less attention has been devoted to PANI preparation by chemical polymerization of aniline, probably because of the well-known poor tractability of the polymer prepared by this route making hard its characterization.

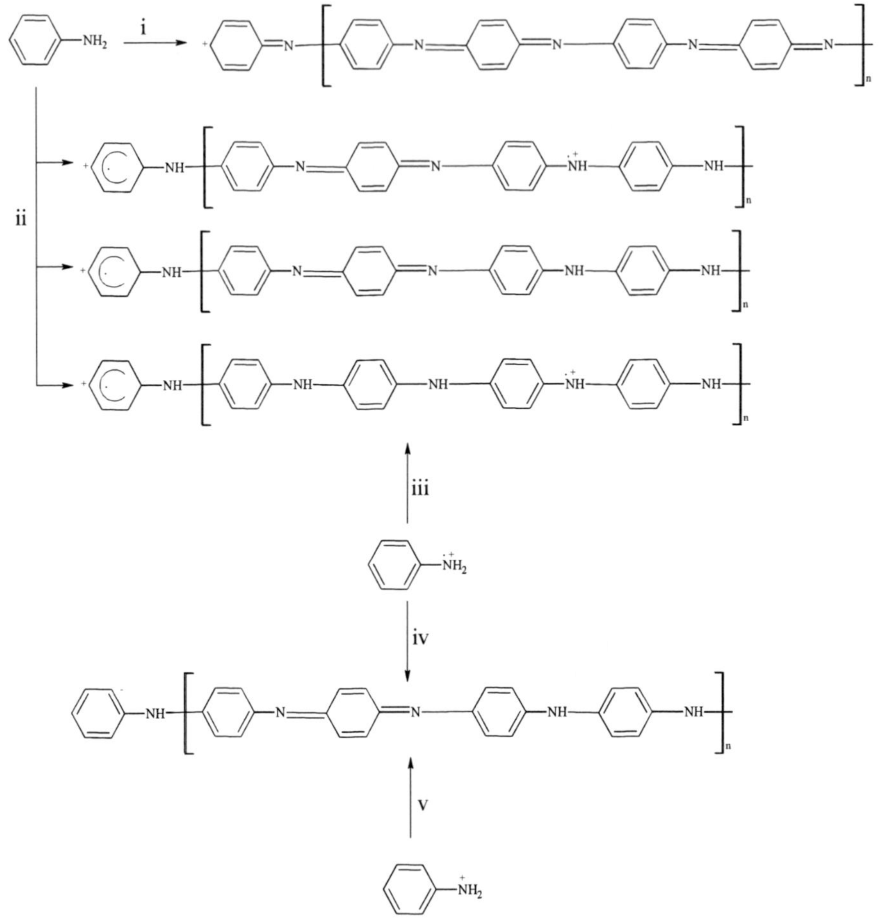

Figure 1.6. Possible reaction pathways for polyaniline growth.

Nevertheless, this latter is still the most popular method to prepare polyaniline in large scale, although it does not allow a fine control of morphology and chemical-physical properties of the polymer.

Even though many researchers sustain that polyanilines prepared by chemical approach or electrochemical approach are undistinguishable both for structure and chemico-physical characteristics, this is not obvious and, even, quite improbable because the chemical process mechanism would involve radicals present in the reaction mixture, whereas these are adsorbed at the electrode surface during the electrochemical process.

In any case, the chemical oxidative polymerization of aniline can be carried out employing many different oxidizing agents: ammonium persulfate, [28] metals in high oxidation states, such as gold, [29] copper, [30] iron, [31] and so on.

Various reaction conditions have been investigated that differ for temperature, pH and reactants quantity. In fact, after an initial period when employing an excess of oxidant was common, the pioneering investigations of MacDiarmid et al. suggested the use of oxidant in stoichiometric amount. [32] At the end of the '80s, the investigation of the reaction conditions effect on the properties of polyaniline prepared by chemical route got started and reached an exhaustive assessment in 1989 thanks to the work of Cao et al. [33]

In particular, they investigated the effect of type and concentration of oxidant, aniline/oxidant ratio, reaction temperature, reaction time, type and concentration of protonic acid and pH of the reaction solution on yield, conductivity and viscosity of polyaniline (measured in solution of concentrated sulfuric acid). All these parameters were shown to affect the polymer characteristics even though in different manner. More in detail, they observed that the kind of oxidant strongly influences yield, conductivity and viscosity of polyaniline, as summarized in Figure 1.7 A-C.

The authors adjusted aniline/oxidant molar ratio in order to operate with a value of $k = 2.5\ M_{an}/M_{ox}M_e$, where M_{an} is the number of aniline moles, M_{ox} is the number of oxidizing agent moles and M_e is the number of electrons necessary to reduce one mole of oxidant.

On the basis of these results it is possible to split the oxidizing agents in three main groups: 1) $(NH_4)_2S_2O_8$ and $K_2Cr_2O_7$, 2) KIO_3, $FeCl_3$ and $KMnO_4$, 3) $KClO_3$ and $KBrO_3$.

Polyaniline prepared with oxidants belonging to the first group is characterized by high values of polymerization yield, conductivity and viscosity. On the other hand, polymers prepared with oxidizing agents of the second group differ only for the smaller viscosity values, except for $KMnO_4$ whose employment leads to materials with low conductivity as well. Finally, worse results in terms of yield, conductivity and viscosity were obtained with oxidants of the third group.

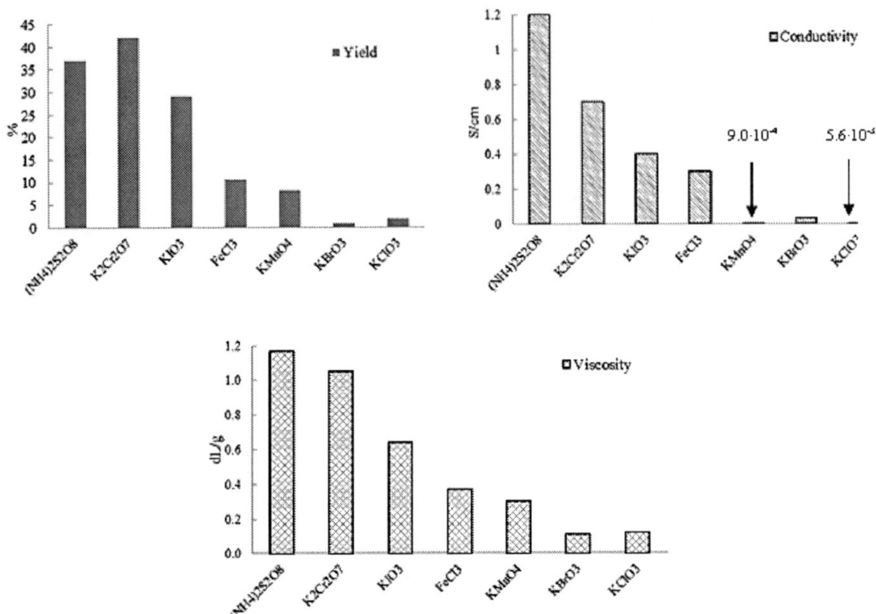

Figure 1.7. A) polymerization yield , B) conductivity and C) viscosity as a function of kind of oxidant. Reaction conditions: aniline/oxidant molar ratio is 2 for $(NH_4)_2S_2O_8$, 6 for $K_2Cr_2O_7$, KIO_3, $FeCl_3$, $KBrO_3$ and $KClO_3$, 5 for $KMnO_4$; reaction temperature= 0°C, [HCl]= 1.5 M.

Aniline/oxidant molar ratio seems to play a key role in the polymerization yield, when using oxidants of the first group (Figure 1.8). No effect has been detected on the other two parameters, conductivity and viscosity.

As far as the protonic acid is concerned, Cao et al. demonstrated that polymerization yield and PANI properties (conductivity and viscosity) are strongly dependent on the type of oxidant employed, leading to better polymers with inorganic acids, such as HCl, H_2SO_4, HBF_4, HF, $HClO_4$, than with organic acids, such as CH_3SO_3H, CF_3SO_3H, CF_3COOH, CH_3COOH.

Moreover, for pH<4 both yield and conductivity remained stable, whereas viscosity decreased with pH.

Figure 1.9 shows that both conductivity and viscosity are not strongly sensitive to polymerization temperature.

In general, it is worth noting that viscosity increases as temperature decreases, reaching the best value at -5°C. The unexpected small decrease at -10°C was associated to a probable aniline hydrochloride salt precipitation.

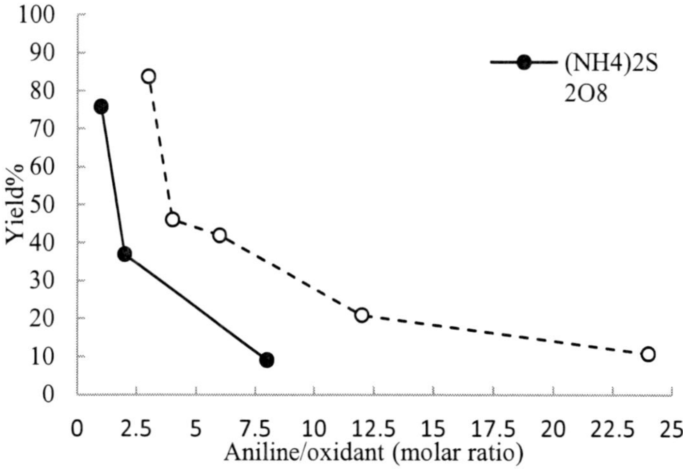

Figure 1.8. Dependence of polymerization yield on aniline/oxidant molar ratio.

Concerning the reaction time, carrying out aniline polymerization at 0°C with ammonium persulfate as the oxidant, a reaction time of 1 h is enough to obtain polyaniline with good viscosity (0.64 dL/g), acceptable yield (41%) and high conductivity (6.1 S/cm). [32]

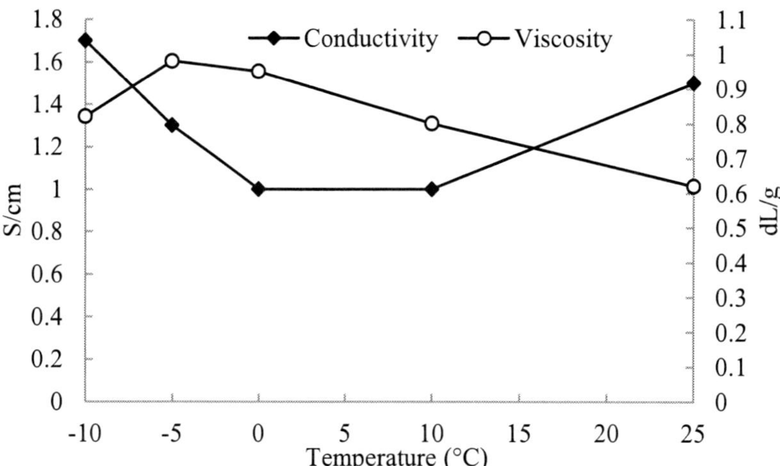

Figure 1.9. Dependence of conductivity and viscosity on polymerization temperature.

In addition to the principal synthetic methods (electrochemical and chemical), more sophisticated approaches have been recently developed, able to produce polymers characterized by appealing properties in terms of morphology, conductivity and solubility. Some of these synthetic pathways are reported below.

1.2.1. Direct and Inverse Emulsion Polymerization

Probably the first attempt to synthesize polyaniline by a direct emulsion process dates back to 1993 by Cao et al. [34]

The addition of a bulky organic acid, such as dodecylbenzene sulfonic acid acting both as surfactant and protonating species, into a reaction mixture containing water, oxidant (generally ammonium persulfate) and aniline leads to a stable emulsion. In this manner, surprisingly, polymers characterized by different morphology and characteristics were synthesized. The authors optimized this process carefully investigating all the parameters (i.e., aniline concentration, aniline/oxidant molar ratio, dopant/aniline monomer molar ratio), thus obtaining a high quality polymer characterized by high molecular weight, homogeneous protonation, high solubility in common organic solvent and fibrillar morphology. They attributed the solubility increase to a higher protonation degree promoted by the reaction conditions.

Highly soluble PANI was obtained by Kinlen et al. in 1998 following a similar synthetic procedure, wherein the heterogeneous system is constituted by water mixed with a water miscible organic solvent (2-butoxyethanol) along with an insoluble organic acid (e.g., dinonylnaphthalenesulfonic acid). [35] The polymer produced according to this protocol was characterized by high molecular weight (Mw > 22,000) and moderate conductivity (10^{-5} S/cm), resulting to be soluble in many organic solvents: toluene, xylene, *m*-cresol, hexane, cyclohexane, chloroform and so on.

Exploiting the higher solubility of the polymer produced by this synthetic approach, PANI-based films were prepared with values of conductivity ranging from $6.5 \cdot 10^{-6}$ to $63.0 \cdot 10^{-6}$ S/cm , as well as ABA triblock films, where A is PANI and B is a diamine terminated derivative of polyethyleneoxide, or polypropylene oxide or polydimethylsiloxane or polyacrylonitrile-*co*-butadiene. [36] The initial conductivity values of the copolymer films (ca. 10^{-5} S/cm) were further enhanced by five orders of magnitude reaching the value of 2.3 S/cm, by treating the surface films with low-molecular weight solvents (alcohols and/or ketones) probably able to remove the less conductive oligomers.

These pioneering studies have stimulated a growing interest in the emulsion polymerization of aniline, leading to conducting materials in form of nanospheres, microsphere, nanotubes with different conductivity by tuning the reaction conditions and mainly the kind of doping agent. [37- 40]

The inverse emulsion polymerization is alternative to the direct method, introduced in PANI synthesis for the first time by Palaniappan and coworkers. [41] PANI/SSA (sulfosalicylic acid) was obtained employing benzoyl peroxide as the oxidant in the presence of sodium lauryl sulfate as the surfactant. Although the product was not soluble in common organic solvents, it displayed a value of conductivity (2.53 S/cm) higher than PANI/SSA achieved in aqueous conditions with ammonium persulfate as the oxidizing agent (0.02 S/cm). Cheaper mineral acids, such as sulfuric, nitric and hydrochloric acid, can also be used. [42] In this case, it was confirmed that sodium lauryl surfactant plays a dual role of surfactant and dopant, as confirmed by X-ray photoelectron spectroscopic (XPS) investigations. Later, Shreepathi et al. reported a new inverse emulsion protocol for the synthesis of soluble PANI using benzoyl peroxide as a mild oxidant, toluene and 2-propanol/water as the solvent and DBSA as dopant as well as surfactant. [43] PANI produced following this pathway resulted to be completely soluble both in chloroform and a 2:1 mixture of toluene and 2-propanol. These organic solutions were used to manufacture films with good adhesive property on metallic and glass substrates. Moreover, in the same paper, the authors clarified the role of aniline/DBSA molar ratio on the polymer morphology. Accordingly, they reported that PANI morphology passes from fibrillar to porous to compact film when DBSA/aniline molar ratio changes from 5:1 to 7:1 to 10:1. Similarly to the direct emulsion polymerization, polymers with different morphology, such as nanoparticles, hollow nanospheres and nanotubes, can be attained by tuning some parameters as the kind and amount of doping agent, surfactant, solvent, oxidant, temperature and reaction time. [44, 45]

Both the direct and inverse emulsion polymerization allow to tune carefully the chemico-physical properties of polyaniline and these methods are still presently the most used synthetic strategies to prepare soluble PANI having specific morphology and good conductivity.

1.2.2. Interfacial Polymerization

Differently from the traditional chemical oxidation wherein all the reactants are closely in contact inside the same reaction medium (generally water), in the interfacial polymerization method aniline and an oxidant are mixed in a heterogeneous system, consisting of two or more different immiscible solvents (generally, water and a non-polar or weakly polar solvent, such as toluene, chloroform or xylene) in the presence of different acids acting as dopants. The interfacial polymerization is closely related to the emulsion polymerization, only differing for the addition of a surfactant able to increase the interface surface area between two liquids having low miscibility.

The interface of two liquids is particularly important for the reactants distribution and this aspect seems to dictate the organization of the oligomers and, hence, of the polymeric chains, thus leading to fibers and, exceptionally, nanotubes. [46-48]

Typically, two solutions are prepared as follows: in the first solution, aniline is dissolved in an organic solvent which can have either a lower or higher density than water; in the second one, the oxidant (i. g., ammonium persulfate) and the acid are mixed. After putting the two solutions in contact, green PANI is immediately produced, initially at the interface and then gradually migrating into the aqueous solution to produce a dark precipitate consisting of nanofibers. The nanofiber diameter depends on the doping acid employed. In 2004 Huang et al. published a paper that definitely clarified why the interfacial polymerization leads to PANI nanofibers, whereas by the traditional approach only PANI with irregular morphology is obtained. [49] They demonstrated that PANI is initially produced as fibers of about 30-35 nm of diameter even by the traditional approach but, when more oxidant is added, these fibers act as template for a secondary growth of polyaniline, initially leading to bigger fibers afterwards to agglomerates with irregular morphology. Figure 1.10 schematically illustrates the difference between the one-phase and the interfacial methods.

1.2.3. Metathesis Polymerization

As demonstrated by Guo et al. the use of aniline monomer for polyaniline preparation is not mandatory, since it can be synthesized even by metathesis reaction, as illustrated in Figure 1.11. [50]

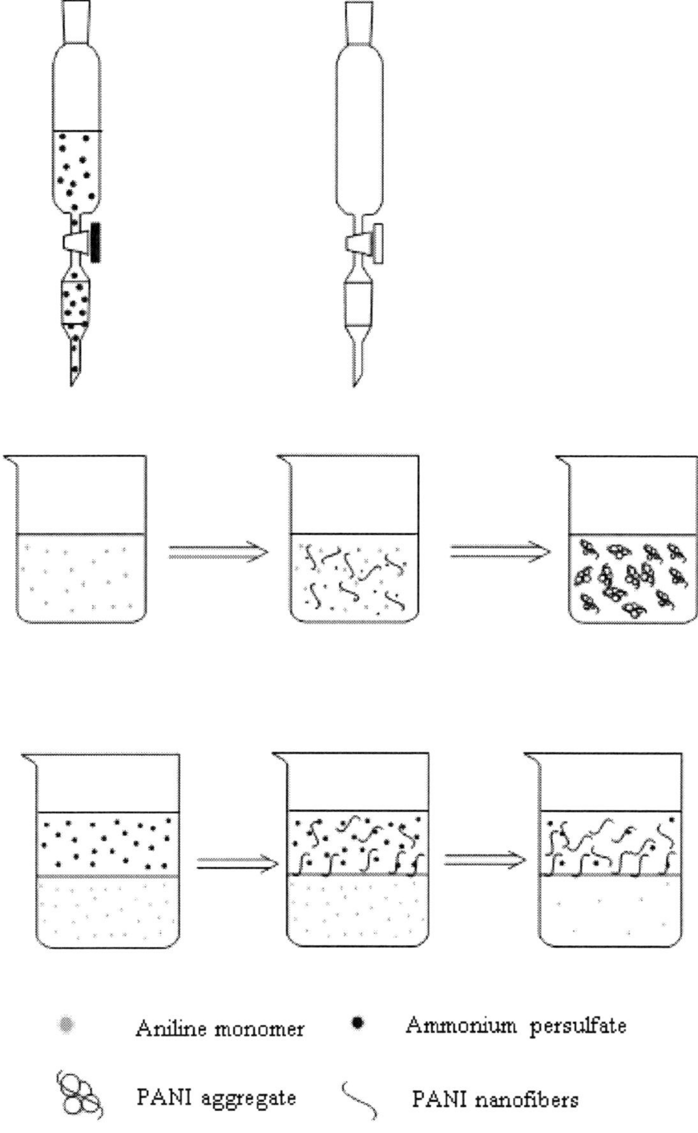

Figure 1.10. Schematic diagram illustrating the formation of polyaniline as agglomerates and fibers during (A) conventional and (B) interfacial polymerization respectively. A) The oxidant (pink circles) is added dropwise into aniline (ochre circles) solution. The initially formed PANI nanofibers aggregate after addition of further oxidant. B) Polymerization reaction takes place at the interface between two immiscible solvents. Fibers produced at the boundary diffuse rapidly into the aqueous phase without any aggregation phenomenon.

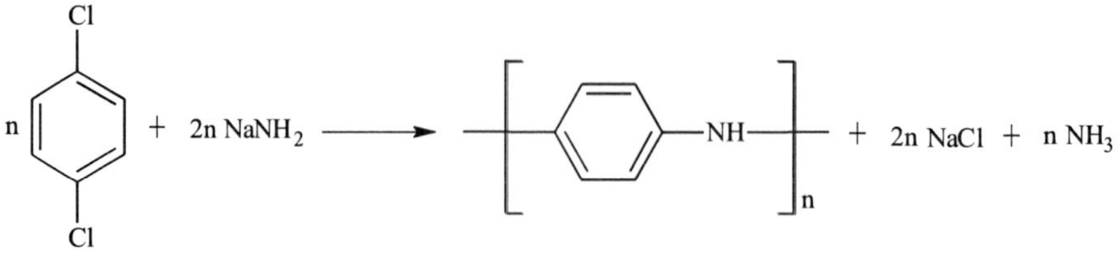

$$n\, C_6H_4Cl_2 + 2n\, NaNH_2 \longrightarrow [-C_6H_4-NH-]_n + 2n\, NaCl + n\, NH_3$$

Figure 1.11. PANI synthesis by metathesis reaction.

The reaction is carried out in a stainless steel autoclave at 220°C for 12 hours using benzene as the solvent. Inspired by the synthesis of poly(*p*-phenylenesulfide) (PPS) commercialized by Phillips

Petroleum Co. in 1973 (PPS has been produced commercially since 1973 by Phillips Petroleum Co. under the registered trademark Ryton), the authors exploited a highly reactive intermediate (Na_2NH) produced by sodium amide ($NaNH_2$) decomposition at high temperature as -NH- source. The subsequent metathesis reaction leads to polyaniline and sodium chloride.

PANI prepared following this approach is not a conducting material. In fact, metathesis reaction is not an oxidative process and the final product is more similar to *leucoemeraldine* than to *emeraldine.*

Although metathesis reaction represents an alternative way to produce polyaniline, this is not competitive with respect to other environmentally friendly routes.

In fact, in order to produce an insulator polymer with modest yield (ca. 50%) a toxic solvent (benzene) is employed. Moreover, the high temperature requested (220°C) entails energy expenditure in disagreement with the modern environmental constraints.

1.2.4. Vapor-Phase Polymerization

The vapor-phase polymerization (VPP) was first described by Mohammadi et al. in 1986 to prepare polypyrrole doped with iron(III) chloride. [51] Later, Bhadra and Lee used this method to synthesize PANI nanofibers with a diameter of 20–100 nm and a length of 50–400 nm. [52] According to this technique vapors of aniline, generated by heating aniline at 200°C,were bubbled into an aqueous HCl solution of ammonium persulfate, the solution kept under stirring for 6 hours and, finally, PANI nanofibers were collected as powder by filtration.

Table 1.1. Solubility of PANI prepared by the vapor synthesis and the conventional method in different organic solvents

	Solubility (g·L^{-1}) in				
Sample	THF	NMP	DMSO	DMF	m-cresol
PANI by vapor polymerization	1.4	5.8	3.2	1.7	3.8
PANI by conventional method	0.5	2.3	1.3	0.08	2.6

The accomplished PANI displayed conductivity values in agreement with PANI prepared following the conventional synthesis (5.8-7.3·10^{-2} S/cm), but with higher solubility (Table 1.1) and thermal stability.

Among all the synthetic methods developed to prepare soluble conductive polyaniline, the vapor-phase polymerization is particularly appealing as it easily leads to nanosized conductive and organo-soluble product by one-pot approach.

1.2.5. Sonochemical Synthesis

The sonochemical synthesis has been widely employed since the '80s. It allows an acceleration of chemical reactions, which can be particularly important in case of heterogeneous systems.

Following this innovative approach, researchers have been able to synthesize not only traditional products in an easier and faster manner, but also new materials not achievable using the existing methodologies. [53]

The power of sonochemical reactions is a physical phenomenon called ultrasonic cavitation.

It occurs when an ultrasonic wave passes through a liquid medium producing a large number of microbubbles, which rapidly grow up and collapse violently thereby producing a shock wave. The bubbles implosion causes locally an enormous concentration of energy expressed as high temperature (ca. 5000 K) and pressure (ca. 2000 atm).

Concerning aniline polymerization, the first examples of PANI prepared following this approach date back to the beginning of this century. These consist in carrying out the reaction in a traditional manner (dropwise addition of oxidant to an acidic solution of aniline) but with the aid of ultrasonic irradiation. Actually, this method allowed Jing et al. to produce PANI NFs with high yield, demonstrating that the further growth and agglomeration of the primary nanofibers are effectively prevented even if an excess of aniline and/or oxidant is used. [54] By the traditional methods polyaniline is intrinsically synthesized as fibers but, if the concentration of aniline and the oxidant is too high, PANI NFs produced in the first step (primary growth) merge very quickly serving as nucleation sites for additional NFs (secondary growth) thereby producing aggregates. One of the most important advantage of this method compared to the traditional ones (especially the interfacial polymerization) is its scalability. This latter aspect makes sonochemical polymerization particularly appealing for the industrial production of PANI NFs.

This fascinating method stimulated many researchers to exploit the power of ultrasonic irradiation for replacing the traditional polluting oxidants with eco-friendly protocols, also employing benign oxidants as hydrogen peroxide. In this regard, following the sonochemical approach and employing H_2O_2 as the oxidant in the presence of Fe^{2+} as the catalyst, Jing et al. accomplished nanofibers of PANI. [55]

1.3. Catalytic Methods: Metal-promoted Oxidative Polymerization Starting from the Monomer (Aniline) or the Dimer (*N*-(4-aminophenyl)aniline)

According to the stringent environmental regulations, the large scale production of conducting polymers, polyaniline included, needs innovative green approaches where eco-friendly oxidants could substitute the traditional and polluting ones. In fact, the biggest drawback in using stoichiometric oxidants is the production of large amount of co-products which must be wasted (1 Kg of $(NH_4)_2SO_4$ per Kg of polymer in the case of ammonium persulfate). Moreover, their presence implies many tedious and expensive steps to obtain highly pure materials. On the contrary, when an oxidative reaction is carried out using more sustainable oxidizing agents, typically H_2O_2 or molecular oxygen, only water is produced as the reduction product, therefore simplifying post-treatments and recycling.

From a thermodynamic point of view, H_2O_2 ($E°= 1.77$ V) would be able to oxidize aniline to PANI. However, this is a low process which can be accelerated by adding a catalyst. Over the years, the oxidative polymerization of aniline and its derivatives has been investigated also under aerobic conditions, that is in the presence of molecular oxygen as the oxidant. Below we report some representative examples of the metal-catalyzed polymerization of aniline.

1.3.1. Cu-promoted Polymerization

Cu-based catalysts are particularly active in the field of oxidative reactions. The first copper-catalyzed polyaniline synthesis was reported for the first time by Toshima et al. in 1994. [56] The extraordinary novelty of their synthetic method was not only the introduction of a catalytic approach but also the use of molecular oxygen as a green oxidant. The authors showed that, among the investigated copper salts, $CuCl_2$ and $CuBr_2$ resulted to be particularly active, reaching TOF values of 520 and 680 h^{-1} respectively. Although the possibility to produce PANI by a copper-catalyzed aerobic oxidative polymerization is fascinating, polymers obtained by this method presented branched structures. Moreover, an accurate spectroscopic investigation revealed that the polymeric structure obtained is more similar to aniline oligormers than to *emeraldine* polymer.

The ability of Cu(II) to catalyze aniline polymerization was later confirmed by Bicak and Karagoz who obtained polyaniline in form of insulating *emeraldine* base by an interfacial polymerization using air oxygen as the oxidant. [30] Dias et al. proposed a more sophisticated catalyst, a copper (II) scorpionate (Figure 1.12), to produce high-quality polyaniline. [57]

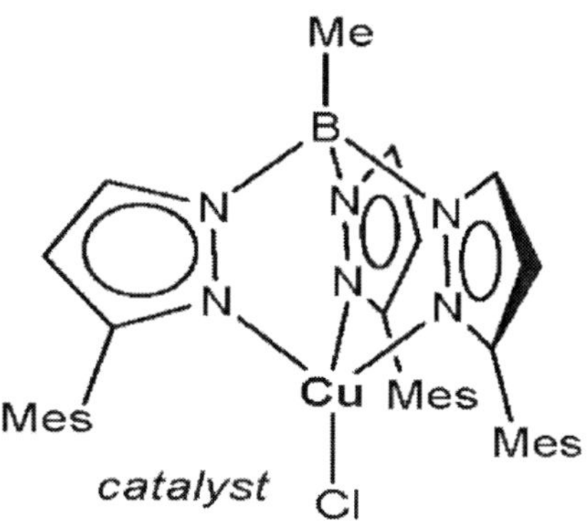

Figure 1.12. Chemical structure of Cu(II) scorpionate [tris(pyrazolyl)boratocopper(II) complex].

The authors reported the oxidative polymerization of aniline dimer (AD), N-(4-aminophenyl)aniline, using hydrogen peroxide as the oxidant and Cu(II) scorpionate as the catalyst. The final polymer was obtained in conducting *emeraldine* form ($2.0 \cdot 10^{-4}$ S/cm) with a modest yield of 50%. However, these results encouraged the authors to carry out further investigations. Hence, they found the possibility to extend this method to aniline monomer thus obtaining polyaniline with high yield (up to 75%) and conductivity (up to $4.5\ 10^{-4}$ S/cm). Despite the exciting accomplished results, the long and difficult process to produce the catalyst, its low solubility in ecofriendly solvents (e.g., water) and the necessity to employ acetonitrile make this catalytic system troublesome for real applications and far from being a green approach.

Inspired by these results and intent to overcome the complication associated to Cu(II) scorpionate, Della Pina et al. investigated the oxidative polymerization of N-(4-aminophenyl)aniline using hydrogen peroxide or molecular oxygen as the oxidizing agents and Cu powder or common copper salts (CuCl and $CuCl_2$) as the catalysts. [58]

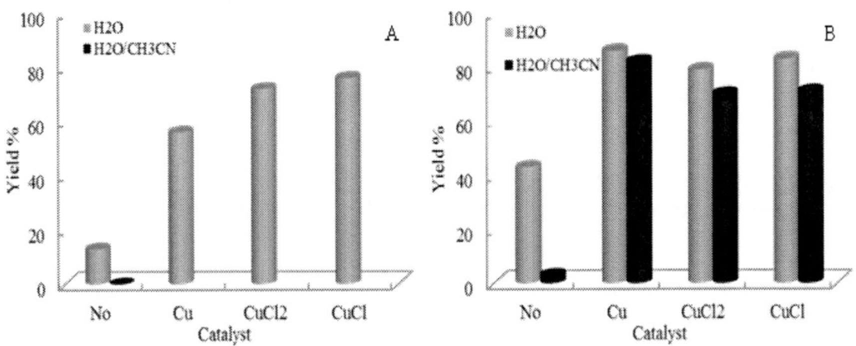

Figure 1.13. Dependence of polymerization yield from the kind of solvent using (A) molecular oxygen and (B) hydrogen peroxide as the oxidizing agents. The reactions were carried out at RT.

They demonstrated that, owing to the lower redox potential of aniline dimer with respect to the corresponding monomer, the dimer could be polymerized using hydrogen peroxide in water also in the absence of any catalyst. Accordingly, PANI could be achieved with 43% yield, whereas employing molecular oxygen as the oxidant led to 13% yield. However, when a mixture of H_2O/CH_3CN instead of pure H_2O was used as the solvent, the yield drastically dropped down from 13-43% to 0-3% respectively in the absence of any catalyst. The strong inhibitory effect of acetonitrile was observed also when the polymerization reaction was carried out under aerobic conditions. Interestingly, the presence of copper catalysts reduced this depressive effect (Figure 1.13).

Finally, the authors investigated the effect of temperature on the polymerization yield under aerobic conditions. Notable 87-93% yields were reached at T = 80°C but, also in this case, a drop from 93 to 77% was registered by acetonitrile addition.

1.3.2. Pd-promoted Polymerization

Even though the preparation of PANI/Pd composites has drawn great attention, only a few authors have reported on the catalytic role of palladium in aniline polymerization.

In 1998, Sadighi et al. prepared controlled-length and functionalized oligoanilines by a simple Pd-catalyzed divergent-convergent and convergent methods. [59] Several strategies have been developed to obtain oligoanilines by polymerization of arylamine or 1,4-phenylendiamine with a protected 4-bromoaniline by a monodirectional or bidirectional growth approaches respectively (Figure 1.14).

Finally a mechanism based on divergent-convergent growth allowed to obtain longer oligoanilines, as described in Figure 1.15.

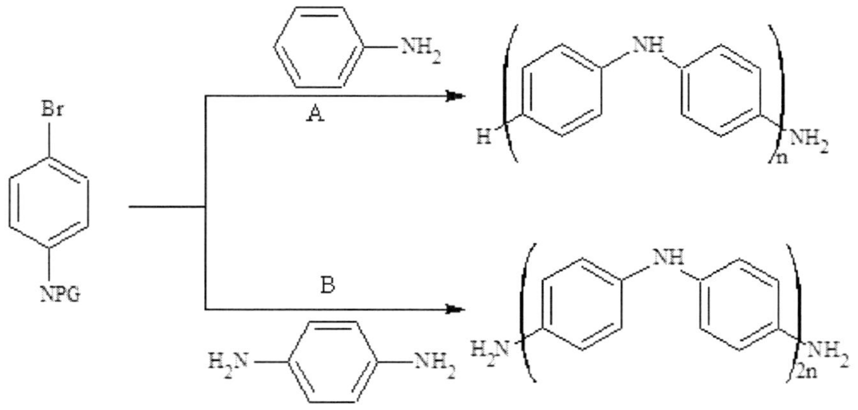

Figure 1.14. (A) Monodirectional and (B) bidirectional growth of oligoanilines by aryl amination. PG= Protecting Group.

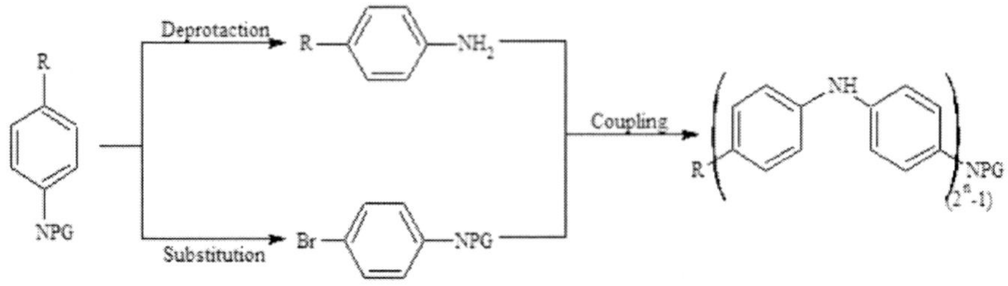

Figure 1.15. Divergent-convergent growth of oligoanilines.

The synthetic method proposed by Sadighi et al. combines the divergent-convergent approach with a modified bidirectional approach obtaining highly symmetric oligomers. Figure 1.16 shows the synthetic strategy suggested by the authors to produce oligoanilines.

Despite this synthetic approach includes numerous reaction steps, it remains a very elegant method to obtain functionalized oligoanilines with a controlled-chain length.

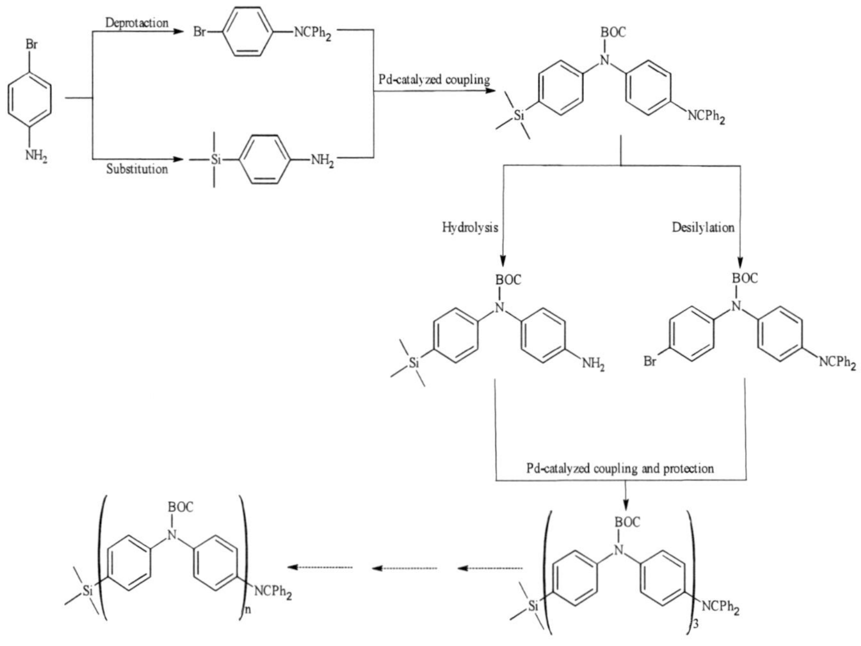

Figure 1.16. Synthetic method for oligoanilines preparation.

More recently, the catalytic activity of Pd nanoparticles in the oxidative polymerization of aniline was discovered. During the synthesis, Pd NPs remain embedded in the polymeric matrix leading to PANI/Pd nanocomposites. [60]

1.3.3. Au-Promoted Polymerization

Gold/polyaniline nanocomposites represent one of the most investigated materials due to their innovative applications in many fields, such as catalysis, sensors, drug delivery, non-volatile memories and so on. [61] The incorporation of gold nanoparticles inside the polymeric matrix has been realized in different manners as it will be discussed in Chapter 2.

Among all the synthetic methods proposed in the literature, only a paper reports the catalytic activity of colloidal gold in aniline oxidative polymerization. [62]

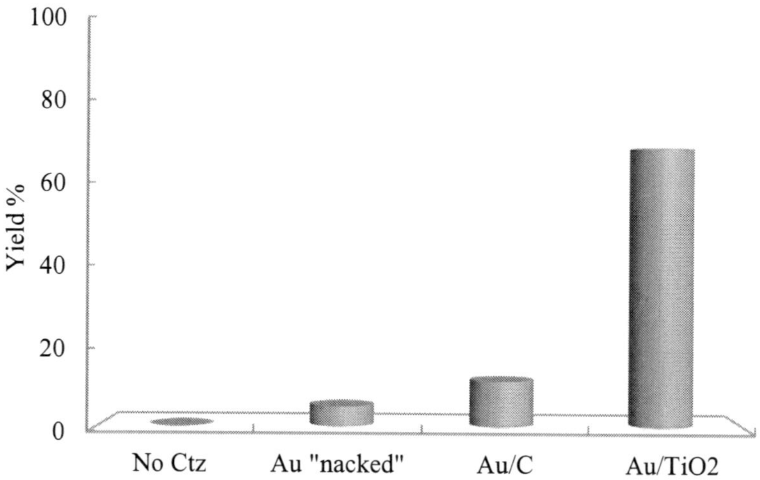

Figure 1.17. Polymerization yield vs kind of catalyst. Aniline/Au= 1000 molar ratio. H_2O_2/Aniline= 1 molar ratio.

The authors investigated the catalytic activity of nano-sized gold (mean diameter of 3.6 nm) as “naked” and supported particles on carbon and TiO_2-P25 in the presence of hydrogen peroxide as the oxidizing agent (Figure 1.17).

As it is possible to observe, PANI is achieved with low yield (about 11.5 %) when "naked" nanoparticles are used as the catalyst. Neither the use of high amount of the oxidant (up to H_2O_2/aniline= 5 molar ratio) nor an increase in the catalyst aliquot promoted PANI production, since only the maximum value of 27.3% yield was reached.

The low catalytic activity of unsupported gold particles could be attributed to their short life-time under the reaction conditions, whereas the more stable supported nanoparticles guaranteed higher performances. In particular, the good results obtained in the presence of Au/TiO_2 motivated the authors to better investigate the reaction. They found that, whereas unloaded carbon resulted to be inert for aniline polymerization, unloaded titania catalyzed the partial aniline oxidation to oligoaniline, identified by the colour change of the reaction mixture from colourless to dark brown.

Moreover, the presence of gold catalyst resulted to affect even the polymer morphology leading to microrods alternating with nanospheres (44-160 nm).

At the best of our knowledge, this still remains the unique example of gold catalysis applied to polyaniline synthesis.

1.3.4. Ag-promoted Polymerization

The pioneering investigations of de Barros et al. opened the way to the preparation of PANI/Ag nanocomposites substituting the traditional oxidants with the combination of photons and metallic silver ions. [63] They observed that, when a solution containing aniline and Ag ions is illuminated with UV radiation (365 nm) or visible light, a characteristic colour change is observed from colourless to green, indicating polyaniline formation, subsequently confirmed by spectroscopic characterization of the obtained polymer.

The catalytic effect of silver ions is confirmed by the fact that the polymerization reaction does not occur in the dark, thereby demonstrating that silver ions are not able to oxidize aniline monomers. Moreover, the reaction time results lower under visible light excitation than under UV irradiation, suggesting that polymerization kinetics is strictly dependent on the excitation energy.

The reaction mechanism proposed by the authors is shown in Figure 1.18.

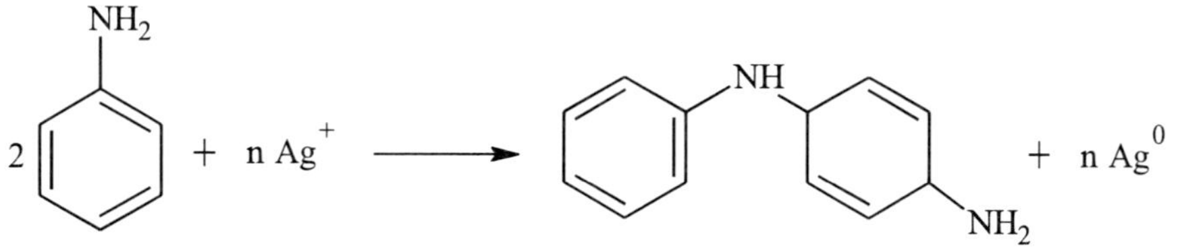

Figure 1.18. Proposed mechanism for PANI/Ag nanocomposites synthesis.

Accordingly, the photo-excitation of aniline monomers produces radical species and the consequent electron transfer to silver ions thus reduced to metal. Aniline monomers can, then, polymerize by a head-to-tail coupling. This method promotes the formation of composites characterized by interesting morphologies ranging from granular PANI containing silver nanowires under UV irradiation to silver-free PANI fibrils under visible light radiation.

Later, Khanna and coworkers re-examined this reaction proposing a new and clearer reaction mechanism. The authors sustained that the key step of the reaction is the simultaneous photo-reduction of silver ions to metal by aniline monomers and the photo-generation of cations, producing silver nanoparticles well protected by aniline and polyaniline deriving from the reaction among aniline monomers cations. [64]

Moreover, silver nitrate resulted to be an excellent catalyst for the synthesis of lignosulfonic acid doped polyaniline (LSA-PANI) in the presence and absence of co-dopants, such as methanesulfonic acid (MSA). [65] In this case, the reaction mechanism remains unclear. In fact, the polymerization reaction is carried out by ammonium persulfate, a stoichiometric oxidant, in the presence of metal ions including silver ions. According to the literature, [66] the addition of a second oxidant (metal ions) reduces the induction period typical of aniline polymerization reaction by persulfate due to the formation of *pernigraniline* as an intermediate in agreement with Figure 1.19

It has been demonstrated that the presence of PANI has a positive effect on the reaction rate, on the contrary an induction period is typically observed for aniline polymerization by persulfate in the absence of polymer. [67]

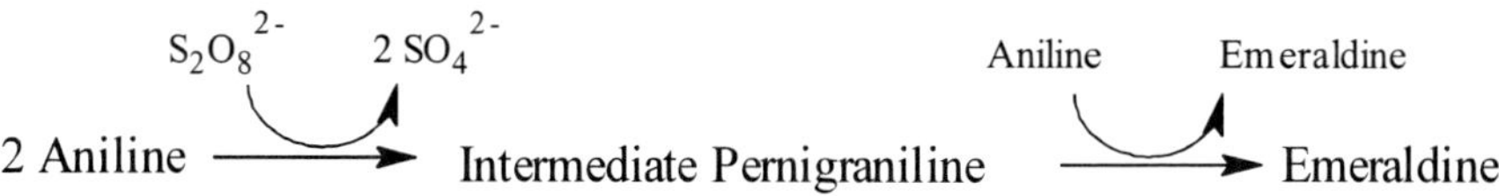

Figure 1.19. Aniline polymerization by persulfate.

In this context, Khanna et al. demonstrated by electrochemical investigations that the metal ions speed up the reaction both in the presence and in the absence of LSA, acting as second oxidizers faster than persulfate in the intermediate *pernigraniline* production. [64] Moreover, they observed that aniline polymerization in the presence of LSA is lower than the corresponding reaction carried out in their absence. The authors suggested that the interaction between anilinium ion and lignosulfonate macromolecules can result in an increased steric hindrance, delaying the monomer oxidation. In addition to this observation, lignosulfonates are well-known radical scavengers which slow the polymerization rate thus reducing the presence of radical species in the reaction mixture. The addition of metal ions, in particular silver nitrate and ferrous sulphate, resulted to be more efficient aniline oxidants thanks to their smaller dimension (Figure 1.20).

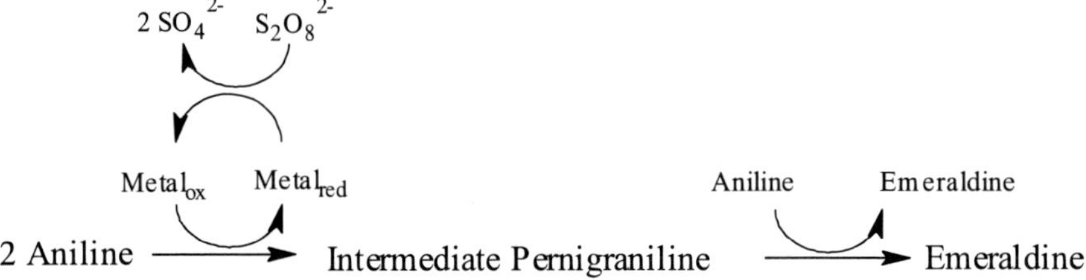

Figure 1.20. Aniline polymerization by persulfate in the presence of metal ions.

At the end of the reaction, silver cations (Ag^+) were definitively reduced as metal particles (Ag^0) and part of them remained embedded in the polymeric matrix leading to PANI/Ag nanocomposites.

Finally, the catalytic activity of Ag nanoparticles (mean diameter of about 6 nm) in aniline oxidative polymerization was investigated by Nadagouda and Varma. In this case PANI/Ag nanocomposites were obtained with a specific morphology of nanorods, particularly appealing for technological and biological applications. [60]

1.3.5. Fe-promoted Polymerization

Iron-based species are probably the most investigated as oxidants and catalysts for polyaniline preparation. In this regard, the first investigation goes back to 1991, when Moon et al. demonstrated the ability of H_2O_2-$FeCl_3$ system to catalyze the oxidation of *leucoemeraldine* to *emeraldine*. [68]

It is known that undoped polyaniline under air actually promotes the oxidation of *leucoemeraldine* to *emeraldine*. In order to investigate more deeply this spontaneous oxidation, Moon and co-workers carried out spectroscopic investigations by molecular oxygen and H_2O_2 in the presence of $FeCl_3$ as the catalyst, and dibenzoyl peroxide.

Initially, the authors followed the UV-vis spectrum variation of a polyleucoemeraldine (PLM) sample dissolved in 1-methyl-2-pyrrolidinone under air, observing a gradual colour change of the solution due to the polymer oxidation related to an increase of the band at 630 nm and a decrease at 345 nm. On the contrary, the polymer solution did not change if stored under inert atmosphere, typically nitrogen or argon. All these results suggested the ability of molecular oxygen to oxidize PLM to *emeraldine* and further investigations demonstrated the promoting effect of light in accelerating the reaction. Despite its higher oxidizing power, H_2O_2 oxidized PLM only in the presence of catalytic amounts of $FeCl_3$ under inert atmosphere in the dark. The reaction was faster than under oxygen and the presence of catalyst was indispensable. Moreover, when carried out under light irradiation, PLM oxidation was slow also in the presence of catalyst, probably for the H_2O_2 photo-assisted decomposition. Generally, the oxidations with H_2O_2 and Fe catalyst (Fenton reagent) lead to the oxidant decomposition producing ·OH radicals which are the real oxidizing agent. However, in this case, if the catalyst and the oxidant are added into the PLM solution simultaneously, the oxidation reaction does not take place. On the contrary, if they are added in two steps, first the catalyst and then the oxidant, PLM is converted into *emeraldine*. This suggests that the active species, produced by the PLM-$FeCl_3$ interactions, are crucial for the reaction.

Along with molecular oxygen and hydrogen peroxide, also dibenzoyl peroxide was investigated as oxidant for the PLM oxidation leading to similar results.

The reaction parameters and mechanism were investigated more thoroughly by Sun and co-worker. [69] They observed that, whereas the catalyst amount is irrelevant for the overall properties and polymerization yield, these latter are markedly dependent on many other parameters, such as temperature, time, aniline and acid concentration as well as initial molar ratio of oxidant to aniline.

For instance, increasing temperature by15°C, the polymerization yield slightly increases whereas the conductivity values drastically fall. The low values of yield observed at low temperature can be attributed to the slow H_2O_2 decomposition under these conditions, thereby reducing the polymerization velocity. Conversely, if temperature is increased up to 20°C, H_2O_2 decomposes more quickly thus speeding up the reaction meanwhile increasing the number of side reactions. This leads to high branching degree of the polymer and reduces its structural regularity. All these factors negatively affect polyaniline conductivity as well. Furthermore, increasing the reaction time influences polymerization yield and conductivity as well as the viscosity of its organic solutions. Since H_2O_2 decomposition increases with time, also yield gradually grows up reaching a constant value of about 66% after 68 h. Viscosity, as well as conductivity, reached the maximum values of 0.34 dL/g and 9.0 S/cm respectively, after 40 h, indicating a high polymerization degree which suddenly decreased probably for hydrolysis and/or degradation phenomena.

Concerning aniline concentration, 0.5-1.0 mol/L resulted to be the best range. In fact, lower concentration values led to higher yields but lower conductivity and viscosity. Similarly, for aniline concentration values higher than 1.0 mol/L, yield remained almost constant but conductivity and viscosity values significantly dropped down.

The acid plays also a fundamental role being crucial for the polymerization to take place. On the one hand it acts as a doping agent for the polymer, on the other hand it increases the redox potential of H_2O_2. In terms of yield, conductivity and viscosity, the best results were obtained for 1.3 M HCl. For values of acid concentration higher than 2 M, not only yield but also other characteristics of the polymer (conductivity and viscosity of the corresponding solutions) decreased. The drop of the material performances is attributed to the chlorination and reduction of quinoid groups.

Along with H_2O_2-Fe^{2+}species, other catalytic systems have been investigated. Among them, Toshima et al. introduced Fe^{3+} catalyst and ozone according to the scheme reported in Figure 1.21. [70]

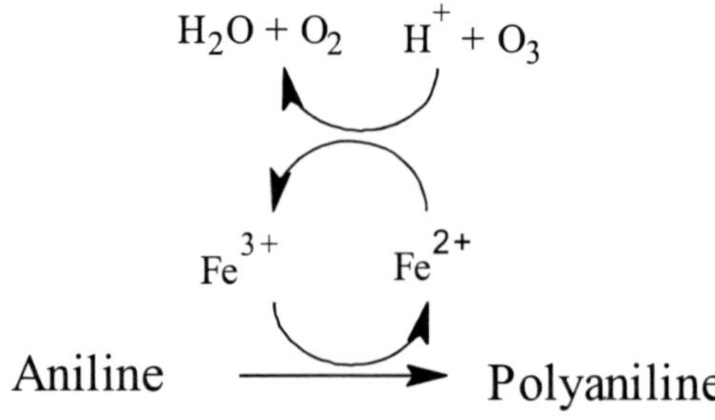

Figure 1.21. Aniline polymerization by Fe^{3+}-O_3 catalytic system.

Owing to its high redox potential (0.77 V), iron (III) could be used as the oxidant to oxidize aniline to polyaniline, while Fe^{3+} was employed as the catalyst in combination with ozone (2.1V), because molecular oxygen has a too low oxidation potential (0.40 V) to convert Fe^{2+} to Fe^{3+}.

Over the years, iron-based catalytic systems were investigated more thoroughly.

For example, the known ability of the enzyme horseradish peroxidase (HRP) to polymerize aniline opened the way to an innovative generation of catalysts based on organic metal complexes. Some of them were enzymes, such as hematin and haemoglobin. [71]

In order to overcome the principal disadvantages of HRP (cost, low catalytic activity and instability in pH < 4 aqueous solutions), synthetic materials were prepared as catalysts.

In this regard, Nabid et al. focused their study on the preparation of metallophorphyrins and metal phthalocyanine, evaluating their catalytic activity in the oxidative polymerization of aniline. [72, 73] The authors identified in the oxoiron (IV) radical cation the species responsible for aniline cation radical formation, which attacks other aniline molecules leading to oligomers and, hence, to polymers.

Thanks to its high stability, the synthetic porphyrin catalyst resulted to be able to carry out the polymerization reaction at various pHs, even though the best results were obtained at pH 4.

Encouraging from these exciting results, Nabid and co-workers extended their investigation to other types of organic metal complexes as the catalysts for aniline polymerization.

In this regard, iron, cobalt and manganese phthalocyanines resulted to be more active, stable and cheaper than the corresponding traditional porphyrines. [73] In particular, iron-based catalyst led to the best results allowing to carry out the reaction in aqueous pH 2 solutions.

More recently, the extraordinary high catalytic activity of Fe_3O_4 nanoparticles as powder and ferrofluid in the oxidative polymerization of *N*-4-(aminophenyl)aniline (aniline dimer, AD) was reported. [74] The main goal of this approach was to produce PANI/Fe_3O_4 nanocomposites with both electrical and magnetic properties under aerobic conditions. In fact, the reaction could be carried out using both H_2O_2 and molecular oxygen as the oxidizing agents, obtaining similar results.

The authors observed that the magnetic nanoparticles were entirely embedded in the final composites, suggesting that the catalytic activity of Fe_3O_4 NPs is not imputable to their dissolution in the reaction mixture.

Although both powder and ferrofluid-type NPs showed similar catalytic activity in the AD polymerization, their effect on the nanocomposites morphology resulted to be different. In fact, concerning PANI/Fe_3O_4 morphology, whereas Fe_3O_4 NPs as ferrofluid preferentially led to nanorods, Fe_3O_4 NPs as powder gave irregular structures.

1.4. The Effect of Doping Agents

Despite three decades have already gone by, the investigations of Chiang and MacDiarmid still represent a milestone, especially for the dependence of PANI conductivity on the protonic acidic doping. [75]

The protonic doping of polyaniline consists in the formation of stable PANI salts. Along with the oxidation level discussed above, the protonation degree plays a key role in the conductivity of this material. The authors investigated the relationship between PANI conductivity and pH of the aqueous solution used for the doping process. These studies allowed to calculate the pKa of the *emeraldine* salt and its dependence on pH.

One might think that owing to their great strength as bases, amine sites would be mainly involved in the protonation process.

However, Chiang and MacDiarmid proposed that imine atoms are preferentially protonated, leading to *emeraldine* salt stabilized by resonance. Due to a high π delocalization which guarantees the characteristic electrical conductivity of polyaniline, all nitrogen atoms are intermediate between amine and imine, all C-N bonds are intermediate between double and single bonds and finally all aromatic rings are intermediate between benzenoid and quinoid.

Concerning the dependence of HCl-polyaniline on pH, at values lower than 4 polyaniline is essentially not protonated. However, for pH higher than 4 the conductivity values grow exponentially reaching a plateau at about pH 0. It is impressive to notice that a variation of only 3-4 pH units causes an increase in conductivity of about 10^{10}.

Regarding the effect of doping on PANI electroconductive properties, the authors observed that after a sudden increase at very low doping percentage (less than 10%), conductivity reached a plateau even when the polymer became highly doped.

The doping percentage was calculated by the equation reported below (Equation 1.1):

$$\%\text{Doping} = \frac{[\text{Cl}]}{[\text{N(tot)}]} \text{ x } 100 \qquad \text{(Eq. 1.1)}$$

where [Cl], corresponding to the number of protonated imine sites, is the number of moles of chlorine determined by elemental analysis, whereas [N(tot)] represents the total number of moles of nitrogen determined by elemental analysis.

These latter data were also used to calculate pKa of the *emeraldine* salt formed in HCl aqueous solutions at different pH.

The *emeraldine* salt dissociation in water can be represented by Equation 1.2:

$$(\text{NH})^{+} \rightleftharpoons (\text{N}) + \text{H}^{+} \qquad \text{(Eq. 1.2)}$$

The pKa for *emeraldine* salt is (Equation 1.3):

$$pKa = pH + log\frac{[(NH)+]}{[(N)]} \qquad \text{(Eq. 1.3)}$$

where $(NH)^{+}$ represents the number of moles of protonated nitrogens, while (N) is the number of moles of non-protonated nitrogens, given by Equation 1.4:

$$[(\text{N})] = [\text{N(tot)}]\text{-}[(\text{NH})^{+}] \qquad \text{(Eq. 1.4)}$$

It follows (Equation 1.5):

$$pKa = pH + log\frac{[(NH)+]}{[N(tot)]-[(NH)+]} \qquad \text{(Eq. 1.5)}$$

The correlation between pKa of *emeraldine* salt doped with HCl and pH of HCl aqueous solution is shown in Figure 1.22.

Emeraldine salt acts as a polyelectrolyte and its pKa varies with the pH of the solution in which it is measured.

The effect of dopant on PANI conductivity was thoroughly investigated by Catedral et al. [12]

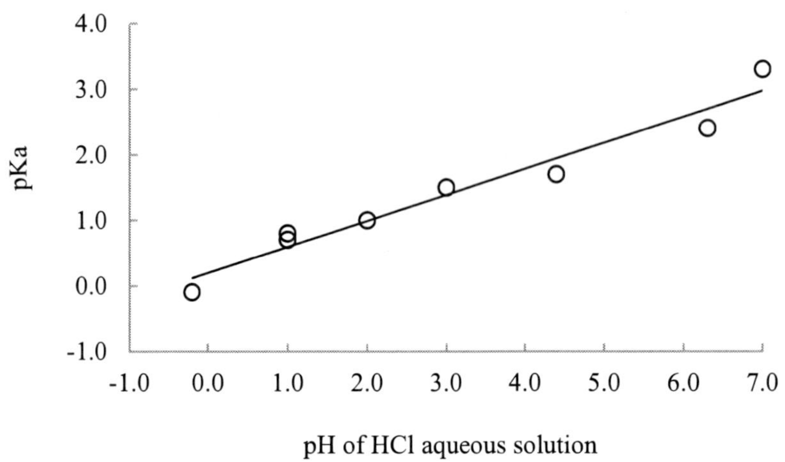

Figure 1.22. Relation between pKa of *emeraldine* salt doped with HCl and pH of HCl aqueous solution.

Comparing the electroconductivity of *emeraldine* base with different PANI samples doped with five inorganic acids (HCl, HNO_3, $HClO_4$, H_2SO_4, HI), they found that the nature of the acid strongly affects the conducting properties.

In particular, $HClO_4$-doped PANI showed the highest values of conductivity (109.04 S/cm), $2 \cdot 10^5$ times higher than *emeraldine* base conductivity. On the contrary, the lowest values were obtained using HI as the dopant (0.02 S/cm). Furthermore, fitting the conductivity values versus computed HOMO-LUMO energy gap data, the authors found an inverse correlation.

1.5. The Effect of Crystallinity

The structure of the polymeric chains markedly influences the electroconductive properties of the polymer. Generally, high levels of crystallinity lead to high conductivity values, because of the increased order of chains which essentially favours the charges mobility among different polymeric chains (*inter*-chain conduction mechanism). [13] X-ray diffraction has emerged as a powerful analytical technique for the investigation of PANI crystallinity (see also Chapter 3).

Basically, polyaniline is an amorphous or, at most, a semi-crystalline material characterized by a broad peak at around 2θ ~ 20°. It was demonstrated that the doping process has beneficial effect on PANI crystallinity which results to be eventually increased (see below).

MacDiarmid and co-workers reported that the ratio between half-width (HW) and height (H) of the X-ray diffraction peak reflects the order of PANI structure. [76] In particular, low HW/H ratios correspond to high structural orders.

The pioneering investigations of Jozefowicz et al. allowed to distinguish two different classes of polyanilines: class I and class II. [77] Class I consists of materials prepared in form of conducting *emeraldine* salt and then dedoped in ammonia solution. For these polymers the amorphous material (EB-I) leads to a semi-crystalline product (ES-I) after protonation which can be converted again in an amorphous polymer by deprotonation (EB-I). Class II consists of materials prepared in form of *emeraldine* base (EB-II) characterized by 50% crystallinity. In this case, at low doping levels no evident changes in crystallinity were observed. However, while the doping level increases a new crystalline diffraction pattern appears, attributable to the doped form of ES-II, whereas the diffraction pattern of EB-II becomes weaker. This latter completely disappears only at very high level of protonation. Moreover, on the one hand the ES-I deprotonation leads again to the amorphous EB-I on the other hand, if ES-II is dedoped, the semi-crystalline structure of EB-II is not restored, but a new amorphous EB-II is obtained. Starting from these results, the authors proposed the unit-cell parameters for all the investigated structures (Table 1.2).

Table 1.2. Crystal structure of emeraldine. deprotonation respectively

Phase	Protonation level (%)	Space group	*a* (Å)	*b* (Å)	*c* (Å)	α^*	V^{**} (Å^3)
EB-I	0	Amorphous	-	-	-	-	-
ES-I	50	P_{21}	7.05	8.60	9.50	97.5°	570
EB-I***	0	Amorphous	-	-	-	-	-
EB-II	0	P_{bcn}****	7.65	5.75	10.20	90°	450
ES-II	50	P_{21221} or P_{c2a}	7.00	8.60	10.40	90°	620
EB-II***	0	Amorphous	-	-	-	-	-

*α indicates monoclinic angle; ** V indicates unit-cell volume; ***EB-I and EB-II from ES-I and ES-II. ****space group of the average structure.

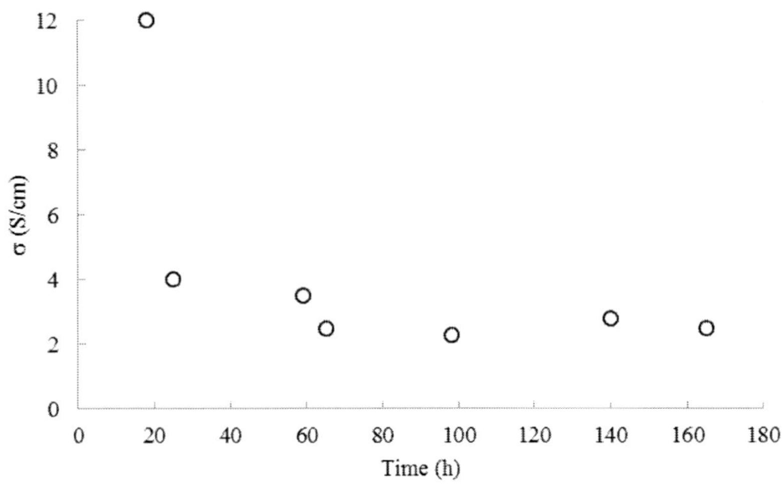

Figure 1.23. Relationship between conductivity and drying time under dynamic vacuum for *emeraldine* salt.

1.6. The Effect of Moisture

The effect of moisture on PANI conductivity has been investigated by many researchers over the years.

Chiang and MacDiarmid, in particular, investigated the relationship between conductivity and drying time under dynamic vacuum of *emeraldine* salt pellets, hence finding that the exposure of polyaniline pellets to water vapour has a positive effect on its conductivity (Figure 1.23). [75]

The authors suggested that water molecules play a role in the anions solvation, reducing the immobilization of positive charges on the backbone and increasing the charge delocalization on the polymeric chains, therefore on the polymer conductivity as well.

In this regard, Focke et al. proposed an elegant mechanism that distinguishes two different contributions in the overall polyaniline conductivity: an intra-chain contribution due to the ability of charge carries to move along the polymeric backbone, and an inter-chain contribution, related to the ability of charge carries to hop between neighbouring polymeric chains. [78]

The two different conduction mechanisms are schematized in Figure 1.24.

In both types of conduction mechanism, water molecules should act as proton transfers, ensuring the charges movement and, hence, increasing their delocalization over the polymeric chains with a positive effect on conductivity.

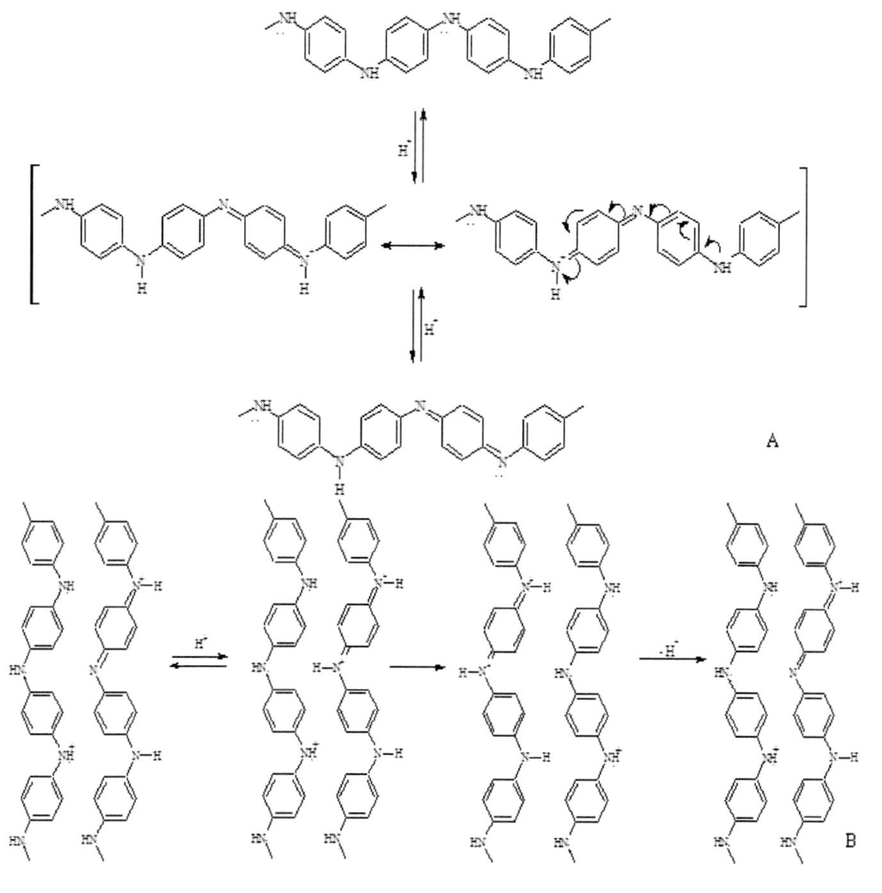

Figure 1.24. (A) Intra-chain and (B) inter-chain conduction mechanism.

In the same period Travers and Nechtschein [15] investigated the diffusion process of water in PANI pellets, finding that the average conductivity (σm) of each pellet is given by Equation 1.6:

$$\sigma m = \frac{4\sqrt{Dt}}{w}(\sigma f - \sigma i) + \sigma i \quad \text{(Eq. 1.6)}$$

where D is the diffusion coefficient of water, t indicates the contact time, W the thickness of the sample, while σf and σi are the steady-state and initial values of conductivity respectively.

They observed that, after setting a given water vapor pressure (ca. 5 Torr), about 24 hours are necessary to reach a steady-state value of conductivity. They interpreted these results as the effect of the diffusion of water molecules into the bulk pellet. Accordingly, they suggested a model where each pellet can be considered as the sum of three slices, two external in equilibrium with water vapor whose conductivity σf is equal to the steady-state value, and one internal whose value of conductivity is the initial σi. The average conductivity σm depends on the ability of water molecules to penetrate inside the pellet reaching the internal slice.

Meanwhile, Angelopoulos et al. performed similar investigations using thin films, rather than compressed pellets, in order to facilitate the equilibration process. [14] They discovered that if polyaniline is used in form of film, the equilibration process is fast. In fact, 90% of the equilibrium conductivity was reached within 20 min in the case of *emeraldine* salt and in 45 min in the case of *emeraldine* base.

On the contrary the desorption process is slower, as 10 hours of exposure to dynamic vacuum were required to recover 90% of the initial resistance.

These results are in agreement with previous investigations that suggested the presence of “metallic islands” inside the polymer. [79, 80] The authors reported that, during the exposure to water vapor, water molecules had a strong effect in decreasing rapidly the inter-particle resistance among the “metallic islands”, meanwhile slowly diffusing into the “islands” thus slightly contributing to the resistance reduction.

In the inverse process (water removal), water molecules among the “metallic islands” were continuously replaced with those contained inside the “islands”, making this process slower than the other one.

Adopting this model, the charges transport in polyaniline is due to their movement among the “metallic islands” and not along the backbone (intra-chain mechanism) or between the polymeric chains (inter-chain mechanism) as proposed by other researchers. [78] Moreover, in this case the limiting factor for the charge transport would be represented by the inter-particles resistance, that is the resistance among the “metallic islands”.

1.7. The Effect of Molecular Weight

The molecular weight or molecular chain length represents another parameter that potentially could affect the conductivity of intrinsically conducting polymers. One would expect that an increase in the molecular weight might positively influence the polymer conductivity but this is not completely true. In fact, very high values of molecular weight are accompanied by several imperfections, such as distortion in chain symmetry, conjugation interruption, local high levels of saturation or unsaturation which adversely affect the charges delocalization and, as a consequence, conductivity. [81] This was confirmed by the research teams of Kitani and Geniès. [82, 83]

A certain value of molecular weight (or polymer size) is necessary so that the long-range delocalization of electron clouds, due to the formation of conjugated double bonds, does not occur until the molecule attains a definite size. The occurrence of long-range charge delocalization is the basis for attaining high conductivity. However, PANI samples characterized by very high values of molecular weight (more than ten times higher than the traditional PANI ones) do not exhibit higher values of conductivity.

Some authors correlated the relationship between conductivity and molecular weight with the two different conduction mechanisms: intra- and inter-contributions. De Gennes and Heeger, for instance, described the conductivity of conducting polymers in terms of τc (the mean lifetime of the charge carrier on the polymer chain) and τi (the time required for a charge carrier to completely explore the polymer chain). [84] Whereas τc mainly influences the inter-chain conduction mechanism, τi has an effect on the intra-chain conductivity. As a consequence, if τc is greater than τI, the polymer conductivity will be affected by the polymer size (or molecular weight). Equation 1.7 shows the linear correlation between polymer conductivity and its chain length.

$$\sigma \propto \frac{ne^2}{k\mathrm{T}} \cdot \frac{aL}{\tau c} \qquad \text{(Eq. 1.7)}$$

where L is the length of the polymer and it is proportional to the molecular weight, a is the persistence length of the chain, n is the charge density and e is the charge per carrier.

On the contrary, if τi is greater than τc, the conductivity will be independent from the polymer size (or molecular weight) (Equation 1.8).

$$\sigma \propto \frac{ne^2}{kT} a\sqrt{(Di/\tau c)} \qquad \text{(Eq. 1.8)}$$

where Di is the diffusion coefficient for the charge carriers along the chain and it is independent from L.

Chapter 2

Synthesis and Properties of PANI-Based Composites

The last decade has seen a growing interest in hybrid electrically conducting nanocomposites. Blending conductive polymers with conventional insulating organic compounds, carbon or graphene, as well as inorganic compounds confer unique properties allowing a wider range of applications than pristine materials. The following sections will present the principal PANI-based composites.

2.1. Reinforcement of Polyaniline via Addition of Carbon-based Materials

Carbonaceous materials, for example, carbon nanotubes and graphene, are generally used as electrodes for capacitors because of their large surface area, porous structure, good thermal and chemical stability, high electrical conductivity and mechanical properties as well as long cycling stability. On the contrary, metal oxides and lightweight conducting polymers turn out to be suitable for pseudocapacitors with relatively high capacitance but they display inferior cyclability due to structural deterioration of the electrodes. Therefore, composites based on appropriate stoichiometric combination of carbonaceous materials and conducting polymers have been proven to be a potential breakthrough for a new generation of efficient supercapacitors and a wide series of novel materials. The reason is the possibility to observe synergistic effects between the two components which give rise to materials with boosted performances. As a consequence, carbonaceous/PANI composites become particularly appealing due to their facile synthesis, processing, as well as the power to synergistically combine the characteristics of carbonaceous materials (i.e., high mechanical strength, high conductivity and large surface area) with the peculiarities of PANI (i.e., wide range of conductivity, doping by oxidation/protonation with insertion of anion).

2.1.1. PANI/Carbon Nanotubes Composites

Carbon nanotubes (CNT), discovered by Iijima in 1991, are one of the four organized carbon states on Earth, with graphite, diamond and fullerenes. CNT are cylindrical shells attained by rolling graphene sheets into a seamless cylinder. Single-wall carbon nanotubes (SWCNT) are made of a single graphite sheet wound on itself with the possibility to close its both ends with a semi-fullerene molecule. Multi-wall carbon nanotubes (MWCNT) consist of a central tube of nanometric diameter surrounded by graphitic layers that are concentrically nested like the rings of a treetrunk. [85] Owing to their unique structural, chemical, mechanical, thermal, optical and electronic properties, CNT and their subsequent exploitation in a wide range of applications are drawing a huge interest. In particular, the field of polymer-based composites represents the most appealing application of CNT. The aim is to transfer the electrical properties of CNT to polymer matrices for obtaining conducting composites or to enhance their pristine conductivity. Combining CNT with conducting polymers, especially PANI, has been observed to allow a synergistic effect between the two components.

The strong van der Waals interactions occurring during the dispersion of CNT make difficult to develop polymer-carbon nanotube composites. A possible solution for facilitating CNT dispersion in organic media would be to functionalize them through non-covalent and covalent reactions with organic molecules. Non-covalent methods allow to preserve the electronic structure of CNT, but the weak forces between wrapped/coupled molecules may lower the load transfer in the composites. [86, 87] On this regard, various protocols are available for carrying out non-covalent reactions between CNT and organic molecules. [88] In general, aromatic structures strongly interact with the basal plane of graphitic surfaces via π-stacking, but a unique compatibility between CNT and aromatic amines, namely aniline, has been detected thus enabling CNT/PANI composites to display good interactions between both the components. As a consequence, the enhanced electrical conductivity of the

CNT-Cl−doped PANI composite, with respect to PANI as such, was explained by the doping effect of CNT owing to a charge transfer from the quinoid unit of PANI to the CNT. Moreover, CNT may serve as a conducting bridge between PANI ES domains. The strong interactions between the two components have been registered also by spectroscopic analyses, as evidenced by the new bands arising at 362, 455, 510 and 550 nm in UV-vis spectrum [89] and the intensity increase of the N = Q = N band at 1137 cm^{-1} in FTIR spectrum. [90] As far as the covalent functionalization is concerned, this may alter the electrical properties of CNT due to disruption of the extended π conjugation in nanotubes differently from the non-covalent methods. PANI chains become less twisted thereby increasing the contact between polymeric redox sites and nanotubes, which leads to more effective electron transport and, consequently, to increase the redox current of PANI. Different methods were developed for achieving amino functionalized CNT which are reactive sites to graft PANI via chemical oxidation polymerization of aniline. [91] Recently, considerable progress has been made in designing and fabricating these composites via electrochemical or chemical routes. [85] The three principal methods to prepare CNT/PANI composite films by electrochemical processing are the one step co-deposition process, the two step process and the electrophoretic deposition process. A comparison between PANI films on nanotubes and traditional electrodes (i.e., platinum) resulted in a superior reducing power of the polymeric film electrodes and, hence, a higher current density from the anodic oxidation. This might derive from the unusual topology and large surface area of the nanotube electrodes.

The chemical routes towards CNT/PANI composites are even more articulated, ranging from the surfactant free aqueous, the interfacial, the plasma and ex situ polymerizations to the micelle–CNT hybrid template directed synthesis or the inverse emulsion pathways. Reporting on each of these protocols is out of the scope of this book, but the reader can find a thorough updated review by Perrin and coworkers. [85]

2.1.2. PANI/Graphene Composites

Graphene is pure carbon whose atoms are assembled to give a honeycomb lattice resulting in an extremely thin, nearly transparent sheet and one atom thick. Hence, graphene is a crystalline allotrope of carbon with 2-dimensional properties, a sort of one-atom thick layer of graphite. Not only it is extraordinary strong despite its very low weight (one hundred times stronger than steel) but it is an excellent conductor of heat and electricity. Although scientists had theorized over it for decades, this material was first isolated only in 2004. Since then graphene (GN) research has boomed but only a decade has passed by, too little time for covering all its potential applications. In carbonaceous PANI-based composites, worth noting is the fabrication of a hierarchical film with coaxial CNT/PANI nanocables, assembled from a complex dispersion of graphite oxide (GO) and CNT/PANI leading to the GO/CNT/PANI precursor film by filtration, afterwards reduced to GN by gas-based hydrazine followed by reoxidation and redoping of the reduced PANI (leucoemeraldine base) to achieve GN/CNT/PANI film. [92] Sheng and coworkers, in their turn, engineered a new hydrogen peroxide biosensor based on enzymatically induced deposition of polyaniline on the functionalized graphene–carbon nanotube hybrid materials. [93]

2.2. Reinforcement of Polyaniline via Addition of Inorganic Compounds

2.2.1. PANI-Au Composites

The growing interest in polyaniline-gold nanoparticles (PANI-AuNPs) composites is related to their notable electrical [94] and biosensing properties. [95] These composites are mainly prepared following three methods: aniline polymerization over preformed Au nanoparticles, [96] Au ions reduction in the presence of preformed PANI [97] and the one-pot aniline polymerization in the presence of gold salt, typically $HAuCl_4$, as the oxidant. [98] Below we report some representative examples concerning method 1.

Feng and coworkers attained PANI/Au composites hollow spheres by addition of an Au colloidal solution to pre-prepared PANI hollow spheres, as schematized below (Figure 2.1). [99]

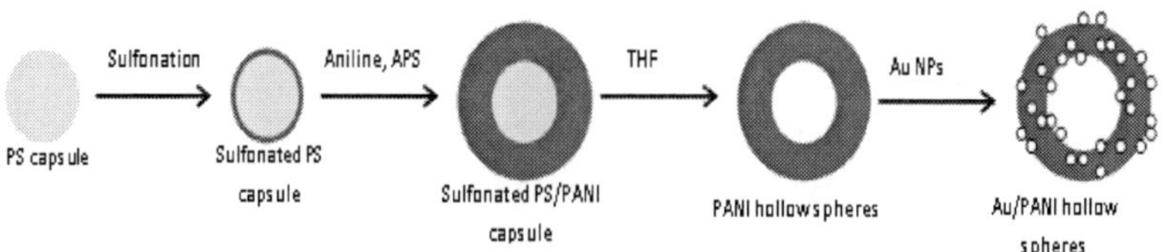

Figure 2.1. Scheme of PANI/Au hollow spheres preparation.

The same authors obtained tubular and fibrillar PANI/Au nanocomposites by a self-assembly method in the absence of any external template. [100]

Kang and coworkers are the pioneers of the second preparation method. [101] They observed that the exposure of polyaniline in both *leucoemeraldine* and *emeraldine* forms to $AuCl_3$ solutions led to the reduction of Au^{3+} ions to Au^0 nanoparticles. On the basis of XPS results, the authors proposed a reaction mechanism where Au^{3+} reduction is followed by polyaniline oxidation and protonation. PANI re-protonation is suggested as the last step of the reaction. The rate of the metal salt reduction and the metal particles size were found to be strongly dependent on the medium used, the initial ratio of metal ions to polyaniline and the reaction time. In particular, when the reaction is carried out in N-methyl-2-pyrrolidinone (NMP), they observed that NPs detection time is strictly related to Au/N molar ratio: the shortest time was observed for the highest Au/N ratio (Figure 2.2).

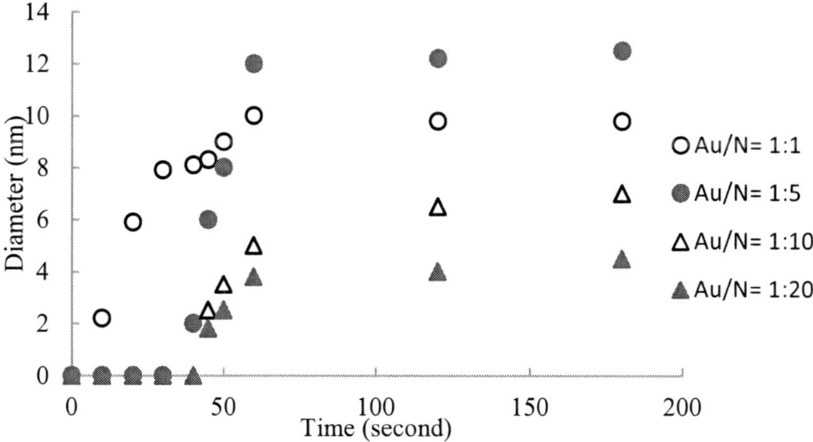

Figure 2.2. NPs dimensions as a function of the reaction time of PANI in $AuCl_3$ in NMP at different Au/N ratios.

After the first 60 min wherein the particle size increased fast, this remained almost constant. Later, they observed that Au^0 NPs size ranged from 4 to 14 nm.

Au NPs growth is time dependent, but it is also correlated to Au/N molar ratio. The largest NPs were observed for Au/N values of 1:1 and 1: 5, even though the maximum diameter value was still <20 nm.

The authors justified this phenomenon considering that the reduction of Au^{3+} ions to Au^0 occurs preferentially on amine groups. [102] Therefore, by decreasing the number of reaction sites (increasing Au/N ratio) gold nanoparticles are concentrated in close sites and agglomeration phenomena will be consequently observed.

When the reaction was carried out in aqueous solution, Au NPs grew bigger, with mean diameter values more than ten times higher than those observed above (Figure 2.3).

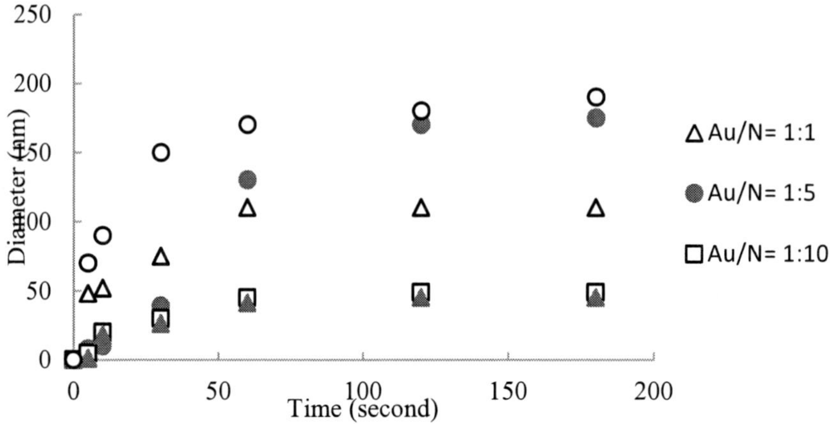

Figure 2.3. NPs dimensions as a function of the reaction time of PANI in $AuCl_3$ in water and in aqueous HCl 1M solution at different Au/N ratios.

In fact, when the reaction was performed in NMP solution, the reduction of Au^{3+} to Au^0 preferentially occurred at nitrogen atoms of the polymeric chains dispersed in solution, whereas in aqueous solution the reduction process interested only the polymeric surface, leading to Au^0 accumulation. Also in this case the highest values of diameter were observed for Au/N ratio of 1:5, whereas increasing the number of reaction sites (nitrogen atoms) and, thereby, reducing Au/N ratio (1: 20), Au NPs reached the maximum value of 30 nm as a mean diameter. However, while retaining Au/N ratio at 1:20 and carrying out the reaction under acidic conditions (HCl 1 M aqueous solution) the diameter grew up to ~ 170 nm. This is attributed to an increase in the reaction kinetics which causes fast agglomeration of Au NPs.

One of the most common method to prepare PANI/Au composites is the aniline oxidative polymerization using gold salts as the catalysts (method 3). One of the most employed gold salts to carry out the reaction is $HAuCl_4$ which leads to PANI nanofibers with mean diameters of 35 nm and length ranging from hundreds of nanometers to several micrometers, as well as gold aggregates of about 26 nm in diameter. [103] As demonstrated by Mallick and coworkers, if the reaction is carried out in organic solution (toluene) gold decorated PANI nanoballs are formed, whereas PANI/Au thin composites films can be attained by interfacial polymerization. [104] Pillalamarri et al. proposed an innovative "grafting form" approach to synthesize PANI/Au nanocomposites. [105] The synthetic method is reported in Figure 2.4.

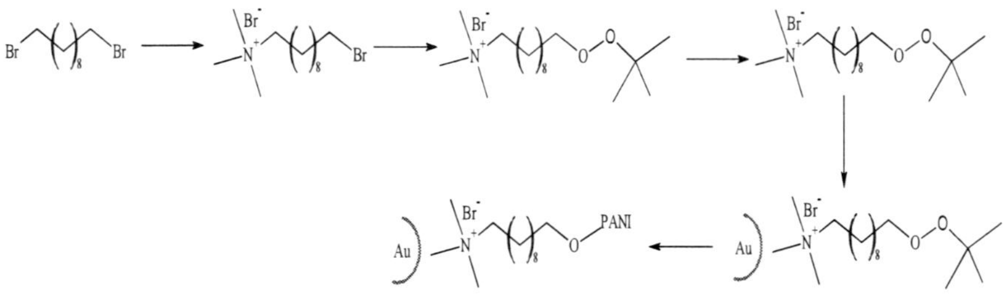

Figure 2.4. Reaction scheme of PANI/Au synthesis proposed by Pillalamarri et al.

Gold NPs, prepared according to a method reported in the literature, were modified by addition of a surfactant containing a peroxide group (10-bromodecylperoxide) able to initiate the polymerization reaction after addition to aniline solution. [106]

Unusual tetrahedral core-shell structures of water soluble PANI/Au nanocomposites were obtained by Peng et al. by the oxidative polymerization of aniline in the presence of $AuCl_4^-$ as the oxidant and sodium dodecyl sulfate as the surfactant to prevent agglomeration phenomena. [107]

In addition to these three main methods, a number of other techniques were developed. Among them, special attention deserves the method reported by Lahav and coworkers allowing to prepare metal-polymer core-shell and segmented nanostructures by a sophisticated electrochemical approach. [108]

2.2.2. PANI-Ti Composites

The combination of polyaniline with extending π -conjugated electron systems and nanostructured

TiO_2 has represented an important prototype in photovoltaics, photocatalysis, photoelectrochromics and sensors. [109] In fact, coupling titania with PANI allows to improve their photoelectrochemical water splitting performance due to a rapid charge separation and slow charge recombination related to PANI delocalized conjugated structures.. [110] A limit in the TiO_2/PANI composite materials so far studied is their particulate form which makes difficult to separate them from the suspensions, whereas films would improve the catalyst recovery. Hence, Zhou et al. recently reported an electrochemical approach for fabricating nanoporous TiO_2/PANI composite films. [111] They prepared nanoporous TiO_2 films on Ti foil following an electrochemical anodic oxidation method, thereafter polyaniline was deposited on TiO_2 surface by cyclic voltammetry. The authors carried out a series of characterization analyses, namely XRD, SEM, Raman and XPS, showing a chemical interaction between PANI and TiO_2. In particular, the authors demonstrated that amorphous PANI has been deposited on the surface of TiO_2, and Ti–O–N–C interfacial bond was formed between TiO_2 and PANI. As a result, the deposition of PANI enhanced the photocurrent from $3.18 \cdot 10^{-5}$ A/cm^2 to $8.15 \cdot 10^{-5}$ A/cm^2 furthermore conferring good stability to the composite films over 3 months. As no enhanced absorption in visible light range was observed after deposition of PANI, the enhanced photocurrent is likely due to a synergistic effect between PANI and TiO_2 which is supposed to promote an efficient separation of photogenerated electron–hole pairs evidenced by EIS. Furthermore, the good stability of TiO_2/PANI composite films in the photoelectrochemical test opens the way to potential applications in photocatalysis and dye sensitized solar cell.

2.2.3. PANI-Fe and -Co Composites

Soft magnetic spinel ferrites with formula $A^{2+}B^{3+}{}_2O_4$, such as Fe_3O_4, $CoFe_2O_4$, $NiFe_2O_4$, $MnFe_2O_4$ and $ZnFe_2O_4$, are widely used in many fields ranging from microwave devices to electromagnetic interference shielding (EMI) due to their high saturation magnetization, high permeability, high electrical resistivity and low eddy current loss. [112, 113]

Iron oxides are naturally widespread compounds, but it is possible to prepare them in various sizes and morphologies also in laboratory. Iron oxides are composed of Fe, mainly in trivalent state, and O and/or OH. Concerning their structure, iron oxides consist of close packed arrays of anions - usually in hexagonal (hcp) or cubic close packing (ccp) - in which the interstices are partly filled with divalent or trivalent Fe principally in octahedral, $Fe(O,OH)_6$ but, in some cases, in tetrahedral, FeO_4, coordination. These various oxides differ for the basic structural units (octahedral or tetrahedral) arrangement in the space.

Materials where Fe^{3+} is the main cationic component are called spinels. There are three main families of ferrites: spinel ferrites, garnet ferrites and hexaferrites, with their own properties different from each other. Concerning the structure of ferrites, oxygens form a fcc sublattice with the cations occupying 16 octahedral (B-sites) and 8 tetrahedral (A-sites) positions. The metal ions distribution is very important to understand the properties of such materials.

Spinels are characterized by direct and inverse structure. More in detail, in a structure of normal (or direct) spinels (AB_2O_4) divalent A(II) ions occupy the tetrahedral voids, whereas the trivalent B(III) ions occupy the octahedral voids in the close packed arrangement of oxide ions. A normal spinel can be represented as: $(A^{II})^{tet}(B^{III})_2{}^{oct}O_4$.

$ZnFe_2O_4$, $FeCr_2O_4$ (chromite) et al. are typical spinels with direct structure. In structures of inverse spinels ($B(AB)O_4$) A(II) ions occupy the octahedral voids, whereas half of B(III) ions occupy the tetrahedral voids. This can be represented as: $(B^{III})^{tet}(A^{II}B^{III})^{oct}O_4$.

Fe_3O_4 (magnetite), $CoFe_2O_4$, $NiFe_2O_4$ et al. are typical spinels with inverse structure.

The number of occupied octahedral sites may be ordered or random. The random occupation leads to defected spinels.

Nanosized particles of iron oxide have emerged as versatile materials for different applications due to their unique magnetic, electronic, photonic and optical properties. The relationship between their structure and function has been intensively studied in magnetic storage, gas sensing, biomedical, and catalysis applications. [114] A wide range of preparation methods are available for achieving nanosized iron oxide, such as sonochemical reactions,[115] mechanochemical synthesis, [116] hydrolysis, thermolysis of precursors as well as co-precipitation technique. [117] In particular, nanoparticles with size distribution virtually near to monodisperse can be produced by thermal decomposition of iron-cupferron complex in octylamine. [118] The synthetic methods leading to polymeric nanocomposites can be distinguished in two types: *in situ* and *ex situ*. The former method is a two-step approach. In the first step the monomer is polymerized in the presence of metallic ions. Otherwise metallic ions can be added at the end of the polymerization reaction. In the second step metallic ions are reduced chemically, thermally or photochemically.

In the *ex situ* method, instead, metal nanoparticles are first synthesized, then passivated and finally dispersed in the liquid monomer which is polymerized to produce a solid nanocomposite.

In both cases the synthesis of the second component is done separately from the monomer polymerization.

Metal/polymer nanocomposites are characterized by peculiar properties which make them particularly attractive for advanced applications (i.e., microwave absorbers, optical filters, materials for photothermal solar collectors, material refractive index ultrahigh/low, materials for magneto-optical and electro-optical).

PANI/metal oxides nanocomposites can be attained following different protocols. Some examples are herein reported. A template-free method has been used to prepare PANI/Fe_3O_4 composites with a core–shell structure, [119] while PANI-Fe_3O_4 nanocomposites through polymerization of aniline in the presence of a ferrofluid has been reported by Kryszewski and Jeszka. [120] PANI composites containing nanomagnets (i.e., Fe_3O_4, d≈14 nm) have been synthesized by a chemical method [121] whereas the solid-stabilized emulsion (Pickering emulsion) route has been followed for achieving PANI/nano-Fe_3O_4 composites. [122] PANI–Fe_3O_4 nanocomposites were obtained through mechanical mixing of dodecyl benzene sulfonic acid doped (DBSA)–PANI powder and HCl-doped PANI–Fe_3O_4 powder. [123] Fe_3O_4/PANI/DBSA with core–shell structure were synthesized by emulsion polymerization. Aphesteguy et al. prepared PANI-Fe_3O_4 through polymerization of aniline in the absence of external oxidant. [124] Mixtures of iron(II) and iron(III) compounds were used as oxidants to polymerize aniline to PANI and form Fe_3O_4 particles in a single step. [125]

Recently, Della Pina and coworkers reported a new clean one-pot synthesis of PANI/Fe_3O_4 composites using O_2 and H_2O_2 as the oxidants and Fe_3O_4 nanoparticles both as magnetic fillers and reaction catalysts under mild conditions. Since MNPs promoted catalytically the reaction of PANI preparation, the effect of nanoparticle size on the synthesis and properties of PANI/Fe_3O_4 nanocomposites was also deepened. [74] Moreover, in order to clarify what is the catalytically active metallic center in the spinel structure, the research group decided to substitute iron(II) with another bivalent cation (Mn, Co, Ni, Cu, Zn, Mg) and to investigate the effect of such a substitution on the catalytic performance. A particular attention has been devoted to the replacement of Fe(II) with Co(II) leading to CoFe2O4NPs as powder (CoMNPsp) and ferrofluid (oleic acid-modified NPs dispersed in toluene, CoMNPsff), thereafter applied in the oxidative polymerization of AD using H2O2 and molecular oxygen as the oxidants.

Although this research is still in progress, it can be anticipated that both CoMNPs powder- and ferrofluid-type resulted to be good candidates as catalysts for the oxidative polymerization of AD. Moreover, differently from Fe_3O_4NPs, the catalytic activity of CoMNPs was higher under aerobic conditions than in the presence of hydrogen peroxide.

2.3. Polyaniline Based Blends: PANI-Insulating Organic Compounds

Despite its extraordinary properties, practical uses of polyaniline are limited by its poor processing, which is strictly related to its low solubility. It is known, in fact, that the only solvents able to solubilize PANI are sulfuric acid or amide-type polar solvents, such as *N*-methyl-pyrrolidinone (NMP), dimethylformamide (DMF), dimethylacetamide (DMAc) and so on. Unfortunately, also in this case only the base form of polyaniline is partially soluble, whereas the doped form, *emeraldine* salt, is intractable and brittle. Many efforts have been addressed to overcome this limitation and some methods have been developed to increase PANI workability. Among them, blending polyaniline with other more soluble polymers is still the most effective, since a host soluble polymer provides mechanical stability and environmental protection of PANI, while the addition of an insulating copolymer reduces the overall conductivity of the material. Sometimes PANI-based blends show problems in terms of miscibility thereby drastically reducing their advantages. For this reason, in order to increase PANI-host polymer miscibility, polyaniline "functionalized" with big organic acids has been employed. [126] Typically, large sulfonic acids are employed as the doping agents, such as dodecylbenzenesulfonic acid (DBSA), camphorsulfonic acid (CSA) and so on, able to improve PANI solubility in common organic solvents.

Other researchers suggested to produce PANI blends employing sulfonated polymers, as polystyrene or polyacrilic acid with the double role of host polymer and doping agent. [127] Various synthetic methods exist to produce PANI-based blends. Typically, they are distinguished in *in situ* syntheses and blending methods.

In situ synthetic methods range from the chemical polymerization/ dispersion of polyaniline in the presence of a host polymer to the electrochemical polymerization of aniline in a matrix covering the anode, as well as the polymer grafting to a PANI surface but also aniline copolymerization with other monomers leading to soluble oligomers.

Regarding the blending methods, PANI is first synthesized by traditional methods and then mixed with other polymers (host polymers). [128] While the first methods (*in situ* syntheses) are cheap and lead to materials characterized by high conductivity, the blending methods are expensive, because PANI cost is higher than aniline monomer, but allow to produce PANI blends in large scale.

Among the host polymers used to prepare PANI blends, poly(methyl methacrylate) (PMMA), poly(ethylene oxide) (PEO) and poly(lactic acid) (PLLA) are the most investigated.

Typically, PANI blends are employed for obtaining films or fibers. Some examples of practical applications are reported in the following sections.

2.3.1. PANI-PMMA Blends

Owing to its low extinction coefficient in the visible light, poly(methyl methacrylate) is one of the most used polymer to prepare transparent conductors if blended with polyaniline. For example, Yang and Ruckenstein synthesized polyaniline/poly(alkyl methacrylate) composites by a two-step emulsion method. [129]

In the first step, an aqueous solution containing aniline and sodium dodecylsulfate (SDS) is mixed with a solution of poly(alkyl methacrylate) in chloroform thus obtaining an emulsion with the appearance of a gel. In the second step, the polymerization reaction takes place by dropwise addition of an HCl aqueous solution of sodium persulfate (oxidant). The blend, precipitated by methanol addition, can be processed by cold- or hot-pressing. Materials prepared by hot-pressing exhibited values of electrical conductivity of about 2 S/cm and good mechanical properties, whereas those prepared by cold-pressing resulted to be more conductive but characterized by poorer mechanical properties.

Many efforts have been addressed to the investigation of charges transport in PANI blends near the percolation threshold. Yoon et al. prepared homogenous films of conducting PANI-CSA-based blends with different sizes and shapes. They reported that CSA-doped PANI networks in PANI/PMMA blends show a very low percolation threshold with a continuous increase of conductivity meanwhile retaining good mechanical properties. [130] Furthermore, the volume fraction of CSA-doped PANI at percolation threshold resulted to reach 0.3% and the conductivity value near the percolation threshold was $3 \cdot 10^{-3}$ S/cm. For PANI-based blends the value of percolation threshold is of crucial importance in terms of transparency. In fact, owing to the high extinction coefficient value of *emeraldine* salt in the visible region, the transparency of its blends can be guaranteed only for extremely low values of percolation threshold. In order to reach these results, Juvin and coworkers proposed to add appropriate plasticizers in PANI/PMMA blends. The presence of plasticizers increases, on the one hand, the blends flexibility and, on the other hand, ensures high values of conductivity and low percolation thresholds. [131] Moreover, the presence of plasticizers reduces the problems related to PANI deprotonation under basic conditions. An interesting result for practical applications was obtained by Hosoda et al. in 2007. [132] They manufactured a conductive paint made of polyaniline/dodecylbenzenesulfonic acid (DBSA) dispersed in PMMA. PANI/DBSA was chemically synthesized in aqueous solution and rapidly extracted by a mixture of toluene and methyl ethyl ketone (MEK) with a ratio 1: 1 v/v. The films exhibited relatively good conductivity and low surface resistivity. When PANI/DBSA and PMMA were mixed with ratio 1:39 w/w, conductivity and surface resistivity resulted to be $9.48 \cdot 10^{-4}$ S/cm and $3.14 \cdot 10^{6}$ Ω/cm^{-2} respectively. These results make such a conductive paint suitable for electrostatic discharge devices.

PANI blends are mainly employed for the preparation of nanofibers and nanotubes by electrospinning technique. This has recently emerged as an effective technique to release well aligned wires and fibers of polymers with diameters ranging from 10 nm to 10 μm which magnifies the intrachain percolation. Three components are required to perform the process: a high voltage supplier, a capillary tube with a pipette or needle of small diameter, and a metal collecting screen. In the electrospinning process a high voltage is used to create an electrically charged jet of polymer solution or melt out of the pipette. Before reaching the collecting screen, the solution jet evaporates or solidifies, and is collected as an interconnected web of small fibers. [112, 133]

Many parameters can influence the transformation of polymer solutions into nanofibers through electrospinning. These parameters include (a) the solution properties such as viscosity, elasticity, conductivity and surface tension, (b) governing variables such as hydrostatic pressure in the capillary tube, electric potential at the capillary tip and the gap (distance between the tip and the collecting screen), and (c) ambient parameters such as solution temperature, humidity and air velocity in the electrospinning chamber. [134]

In order to prepare nanofibers or nanowires, the polymer must be dissolved in a solvent or melted to be introduced in the pipet or syringe. However, it is known that some COPs, especially polyaniline and polypyrrole, are characterized by very low solubility and decompose before to melt.

Therefore, the drawbacks related to poor solubility, low viscosity and surface tension in common solvents must be overcome otherwise PANI cannot be directly processed. Doping polyaniline with appropriate organic acids such as sulfonic acids (i.e., camphorsulfonic acid, CSA, and dodecylbenezensulfonic acids, DBSA) has been found to enhance solubility, whereas adding a copolymer (generally an insulator as polystyrene, polyacrylonitrile, polymethylmethacrylate, polyvinylpyrrolidone and polyethylenoxide (PEO) lowers both viscosity and surface tension thus improving processability.

In this case, in fact, the possibility to have a soluble or a molten material is essential. In this regard, PMMA along with PEO are the most common polymers indicated for preparing conductive blends. Even though many papers have been published on PANI/PMMA nanofibers electrospun preparation, the following examples deserve special attention. Attout, Yunus and Bertrand prepared aligned PANI/PMMA NFs and nanotubes (NTs) by electrospinning technique employing PMMA as a template. [135] More in details, electrospinning of PANI NTs was realized in three steps. The first step employed a hydrophobic silicon surface for enhancing fiber adherence and enabling electrospinning. Thereafter (second step), aniline was polymerized on PMMA fibers by dipping these latter in an aqueous solution containing aniline, H_2SO_4 and $(NH_4)_2S_2O_8$ at 0°C for 1, 2 or 5 hours long. In the third and last step, PMMA core was removed by dissolution in $CHCl_3$ or by heating at 150°C. The thickness of PANI NTs walls (~ 100 nm) resulted to be slightly dependent on the reaction time. Thinner walls were achieved by reducing the immersion time. Moreover, the mean diameter was found to be affected by PMMA fiber size. To produce well aligned PANI/PMMA NFs, the authors employed a sophisticated experimental setup (Figure 2.5).

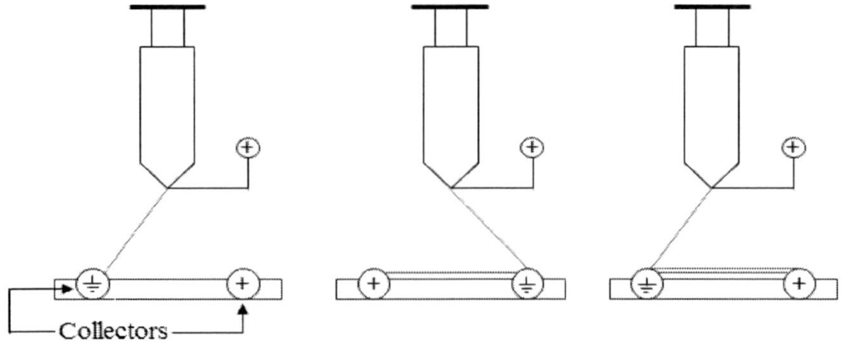

Figure 2.5. Electrospun fibers alignement setup.

The fibers are electrospun and aligned between two collectors, the first one is grounded whereas the second one is maintained at high voltage (6 kV). Moreover, after exit from the needle, the electrically charged polymeric jet is bounced from a collector to another one because of the electric alternate field between the two collectors and, as a result, well aligned nanofibers are produced. PANI/PMMA blend nanofibers exhibited diameters of 150 nm. Owing to the low amount of PMMA in the blend (5%wt) necessary to increase PANI content in the final fibers, the low viscosity of the solutions led to defects (as beads) in the NFs and phase separation related to the capillary instability of the spinning jet by surface tension.

More recently, Zhang and Rutledge prepared 100% PANI/CSA NFs by coaxial electrospinning of CSA-doped PANI/PMMA blended solutions and subsequent stretching along their fiber axis. Then, the copolymer was removed by composite NFs dip in isopropanol solution. [136] After stretching, 100% PANI/CSA NFs exhibited values of electroconductivity up to 130±40 S/cm. The authors attributed such a notable conductivity to the enhanced molecular orientation arising from extensional deformation during electrospinning process.

2.3.2. PANI-PEO Blends

PANI/PEO blends preparation and use are almost exclusively correlated to nanofibers production by electrospinning technique. Conducting polymers in nanofiber form are drawing great attention both from academia and industry due to their potential applications in electronics, textiles and biomedical devices. Nanowires and nanofibers can be produced by various chemical and physical methods, including drawing,[137] template synthesis,[138] phase separation,[139] self-assembly,[140] electrospinning.

In the case of electrospinning, promising results have been achieved. In particular, ultrafine fibers of CSA-doped PANI blended with PEO were prepared by Norris and coworkers. [141] All the NFs were obtained with a mean diameter of about 950 nm and length up to 2.1 μm.

If compared to films, the values of electroconductivity of non-woven PANI/PEO fibers resulted to be low, because of the high porosity of these materials.

Isolating single fibers of 100% H_2SO_4-doped PANI (mean diameter ca. 139 nm), the same authors measured values of conductivity up to 0.1 S/cm. The use of PANI blended with another polymer allows to avoid corrosive compounds. MacDiarmid and coworkers reported values of conductivity of about 10^{-1} S/cm for CSA-doped PANI/PEO (50% w/w) fibers having mean diameters between 419 and 600 nm. [142] These values grew up to 33 S/cm when PANI/CSA content in the blend was increased up to 70% and applied voltage was kept constant (10 mV) to minimize heating effects during the conductivity measurements.

Well oriented CSA-doped PANI/PEO nanofibers producing continuous nanofilaments (mean diameter = 100 nm) were achieved using a rapidly rotating collector. Also in this case the results in terms of electroconductivity were surprising (33 S/cm). [143]

The presence of an insulator copolymer, however, limits the material conductivity. Hence, the possibility to produce pure PANI nanofibers is very tempting. This would allow to obtain materials characterized by high electrical properties while overcoming possible problems of biocompatibility for medical and biomedical applications. It has been, in fact, recently demonstrated that pure PANI is biocompatible and not cytotoxic. [144]

An innovative way towards pure PANI electrospun nanofibers has been reported by Della Pina et al., allowing to reduce the copolymer (PEO) amount from 1 to 0.1% w/w with respect to the amount of PANI doped with dodecylbenzensulfonic acid (DBSA). [145] They found that reducing the amount of PEO in PANI/PEO blend organic solution led to decrease the fiber size from 421 nm, for higher content of PEO (PANI: PEO= 1: 1, w/w), to 230 nm (PANI: PEO= 1: 0.1, w/w) while increasing conductivity from $2.73 \cdot 10^{-4}$ S/cm to $5.33 \cdot 10^{-3}$ S/cm.

Moreover, the way to collect NFs was found to affect their morphology. Accordingly, the use of a rotating collector favors fibers stretching, therefore promoting a narrow distribution of diameters with respect to using a static collector.

2.3.3. PANI-PLA Blends

Recently, poly(lactic acid) (PLA) has emerged as a good alternative to traditional petrochemical-derived polymers for many applications, being a biodegradable and thermoplastic polyester derived from renewable resources. Thanks to its characteristics, it finds application alone or blended with other polymers as a basic material in many fields, such as packaging, building, agriculture, transportation, furniture, tissue engineering and so on. [146]

Combining the above mentioned PLA properties with those of conducting polyaniline, some researchers produced PANI/PLA nano-sized blends. [147] It was demonstrated that PANI and PLA form stable and homogeneous blends which can be exploited to produce nanofibers by electrospinning technique.

In this regard, PANI/PLA nanofibers were easily electrospun first by Picciani and coworkers in 2009 [148] and, one year later, by Rahman et al. [149] More in particular, the former authors fabricated NFs with diameters ranging from 100 nm to 1 μm and very modest conductivity values (up to 10^{-9} S/cm), whereas the latter authors achieved not only polyaniline but also poly(aniline-co-m-aminobenzoic acid) (P(ANI-co-m-ABA)) with poly(lactic acid) (PLA) nanofibers. PANI/PLA NFs were obtained with diameters of about 141 S/cm and the mean diameter of poly(P(ANI-co-m-ABA))/PLA NFs was 124 nm. In both cases, the results in terms of conductivity were very modest: 10^{-6} mS/cm in the first case and 2.0×10^{-5} mS/cm in the second case. Gizdavic-Nikolaidis et al. addressed their efforts to the preparation of HCl-doped poly(aniline-co-3-aminobenzoic acid) (3ABAPANI) copolymer and poly(lactic acid) (PLA) blended together and electrospun. [150] The resulting nanofibrous blends exhibited surprising bioactive properties and good ability to promote proliferation of COS-1 fibroblast cells. Moreover, nanofibrous PANI-based materials showed a positive effect in the growing cells and potential antimicrobial capability against *Staphylococcus aureus* as well as electrical conductivity. In order to increase PANI solubility and reach their scope (conducting fibers production), the same authors did not use the typical method based on PANI protonation with sulfonic acids but they copolymerize aniline with substituted anilines thus imparting solubility to the resulting functionalized PANI copolymers. Concerning this, aniline was copolymerized with aminobenzoic acids (ABAs), obtaining copolymers soluble in basic aqueous media and in polar solvents such as N-methyl-2-pyrrolidone (NMP) and dimethyl sulfoxide (DMSO), easier to be electrospun. Nanofibrous blends exhibited honeycomb structures and conducting properties, useful for electrical stimulation of damaged tissue. All these features make PANI-based nanofibrous blends excellent materials as biocompatible scaffolds for applications in the field of tissue engineering for the production of antimicrobical wound dressings in the absence of any antiseptic.

PANI/PLA blends can be also employed for many other kinds of application. Probably the most interesting was recently reported by Peng and coworkers and consists in the use of PANI/PLA composite nanofibers as counter electrodes for dye-sensitive solar cells (DSCs). [147] The interest in this field is strictly related to the need to substitute the traditional DSCs counter electrodes based on platinum which imply hard conditions for their preparation. This reason, along with the high price and scarce availability of Pt, increases the cost of fabrication thereby limiting their application in large scale. In this context conducting polymers, and in particular polyaniline, might be a good alternative. PANI/PLA composite nanofibers were directly electrospun on rigid fluorine-doped tinoxide (FTO) and flexible indium tin oxide-coated polyethylene naphthalate (PEN) substrates for producing three-dimensional porous materials. Such composites exhibited good photoelectrical conversion efficiency, approaching the performance level of a traditional Pt electrode. These preliminary results, along with the simple and cheap preparation method, make PANI-based composites promising counter electrode candidates for efficient DSCs.

Chapter 3

CHARACTERIZATION OF POLYANILINE AND ITS COMPOSITES: THE MAIN ANALYTICAL TECHNIQUES

With its focus on the characterization of polyaniline and its composites using such techniques as X-ray diffraction and spectrometry, light and electron microscopy, thermogravimetric analysis, as well as cyclic voltammetry, this chapter helps the reader to correctly choose the proper analytical technique to characterize the materials prepared. Each section, in fact, introduces a specific characterization method thus eventually providing a spectrum of the principal analytical tools allowing to predict the microstructure of the synthesized polymer, its thermal and mechanical properties, and so assess its suitability for a particular application.

3.1. ULTRAVIOLET-VISIBLE SPECTROSCOPY (UV-VIS)

This analytical technique refers to absorption spectroscopy or reflectance spectroscopy in the ultraviolet-visible spectral region (approximately,100-800 nm wavelength, λ). This means it uses light in the visible and adjacent (near-UV and near-infrared [NIR]) ranges. The absorption or reflectance in the visible range directly affects the perceived color of the chemicals involved. In this region of the electromagnetic spectrum, molecules undergo electronic transitions since molecules containing π-electrons or non-bonding electrons (n-electrons) can absorb the energy in the form of ultraviolet or visible light to excite these electrons to higher anti-bonding molecular orbitals. The more easily excited the electrons (i.e., lower energy gap between the HOMO and the LUMO), the longer the wavelength of light it can absorb.

One of the major limitations of conducting polymers, in particular polyaniline and polypyrrole, is their very low solubility. Polyaniline is slightly soluble only in a few solvents, as pyrrolidones and amides. A typical UV-vis specimen is prepared by dissolving a minimal amount of PANI in 2-pyrrolidone or *N,N*-dimethylformamide thus obtaining a blue to green solution. In a typical spectrum of polyaniline the band at 320 nm corresponds to the $\pi \rightarrow \pi^*$ transition of the benzenoid rings. The peak at 410 nm corresponds to the polarone-bipolarone transition and the broad peak at 800 nm to the $\pi \rightarrow \pi^*$ transition of the imine-quinoid group. UV-vis spectroscopy allows to investigate how the spectrum of polyaniline varies when leucoemeraldine is oxidized to emeraldine. In fact, the peak at 800 nm increases with the oxidation of the polymer. [151] Using such a technique it is possible to investigate also the behaviour of both leucoemeraldine and emeraldine when an acid as $HClO_4$ is added into the organic solution under air or N_2. While recording the spectra at different times, it has been observed that leucoemeraldine and emeraldine in their protonated forms are metastable in *N*-methylpyrrolidinone.

Accordingly, the aminic nitrogen atoms of leucoemeraldine are deprotonated by the solvent and oxidized by the atmospheric oxygen at the same time, with an increment of the 650 nm band. This does not happen when the specimen is kept under nitrogen. Emeraldine behaves differently. In fact, under air, PANI undergoes deprotonation without any further oxidation. However, in the absence of any oxidant, the oxidation state decreases as the deprotonation goes on. [151]

Moreover, by this powerful technique, MacDiarmid et al. could demonstrate that the oxidation of polyanilines can range anywhere from y = 0 to y = 1, according to the general structure of PANI reported in Figure 3.1.

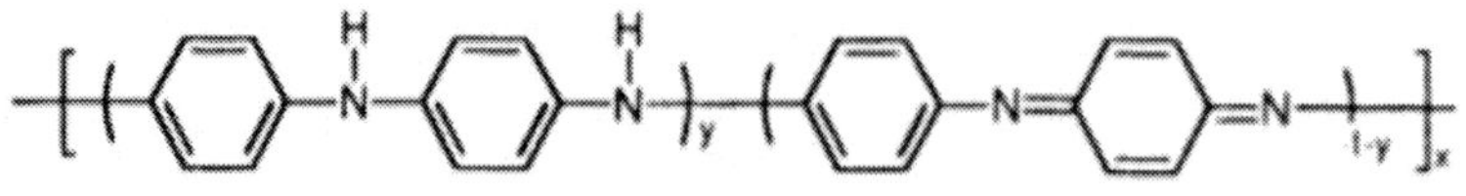

Figure 3.1. General formula for PANI.

In the range from y = 0.5 to y = 1 (from emeraldine to leucoemeraldine form) on a molecular level and in *N*-methyl-2-pyrrolidinone solution only two chromophores are present, characteristic for y = 0.5 and y = 1 species. At a molecular level, all the intermediate oxidation states consist only of mixtures of these characteristic chromophores. [5]

UV–vis–NIR (Ultra Violet – Visible – Narrow Infrared spectroscopy) adds further information on the conformation of the molecules both in solution and solid state. Although PANI in base-form shows a typical coil-like structure, the protonation process is accompanied by the creation of positive charges on nitrogen atoms whose reciprocal repulsions cause straightening of the chains (structure rod-like). This is beneficial for delocalizing the electrons (negative charges) along the chain, therefore creating energetically favorable polaronic structures. The polymer conformation can be strongly influenced by the dopant, as well as the solvent, employed during the synthesis. For example, bulky anionic dopants cannot quickly diffuse in between the chains following the delocalized positive charges and the chain expansion will be hindered. Xis et al. investigated the effect of solvent in the conformation of PANI-HCSA (camphorsulfonic acid) demonstrating that coil-like conformation is predominant in *m*-cresol, *p*-cresol, 2-chlorophenol, 2-fluorophenol and 3-ethylphenol, whereas rod-like conformation in chloroform, NMP (N-methylpyrolidinone), DMF (N, N-dimethylformamide), and benzyl alcohol. [152]

The conformation of polyaniline chains can be determined on the basis of its UV–vis–NIR spectra.

The band extending beyond 800 nm towards the near infrared region is called free-carrier tail and is characteristic for the highly conductive protonated PANI in rod-like conformation. Differently, this effect disappears when protonated PANI is in coil-like conformation.

3.2. Fourier Transform –Infrared Spectroscopy (FT-IR)

Infrared spectroscopy (IR spectroscopy) deals with the infrared region of the electromagnetic spectrum, that is light with a longer wavelength and lower frequency than visible light. It covers a range of techniques, mostly based on absorption spectroscopy. The infrared portion of the electromagnetic spectrum is mainly divided into three regions: the near-, mid- and far- infrared, named for their relation to the visible spectrum. The higher energy near-IR, approximately 14000–4000 cm^{-1} wavenumber (λ = 800–2500 nm) can excite overtone or harmonic vibrations. The mid-infrared, approximately 4000–400 cm^{-1} (λ = 2500–25000 nm) may be used to study the fundamental vibrations and associated rotational-vibrational structure. Finally the far-infrared, approximately 400–10 cm^{-1} (25000–1000000 nm), lying adjacent to the microwave region, has low energy and may be used for rotational spectroscopy. The names and classifications of these subregions are conventions, and are only loosely based on the relative molecular or electromagnetic properties. Fourier Transform-Infrared spectroscopy (FT-IR) is used to obtain an infrared spectrum of absorption, emission, photoconductivity or Raman scattering of a solid, liquid or gas. An FT-IR spectrometer simultaneously collects spectral data in a wide spectral range. This confers a significant advantage with respect to a dispersive spectrometer which measures intensity over a narrow range of wavelengths at a time. The term originates from the fact that a Fourier transform (a mathematical process) is required to convert the raw data into the actual spectrum.

FT-IR spectroscopy is probably the most useful technique for characterizing polyaniline since it allows to easily determine the oxidation degree of the polymer. Figures 3.2 (A-C) reports characteristic Fourier-transform IR spectra for *leucoemeraldine, emeradine* and *pernigraniline* respectively.

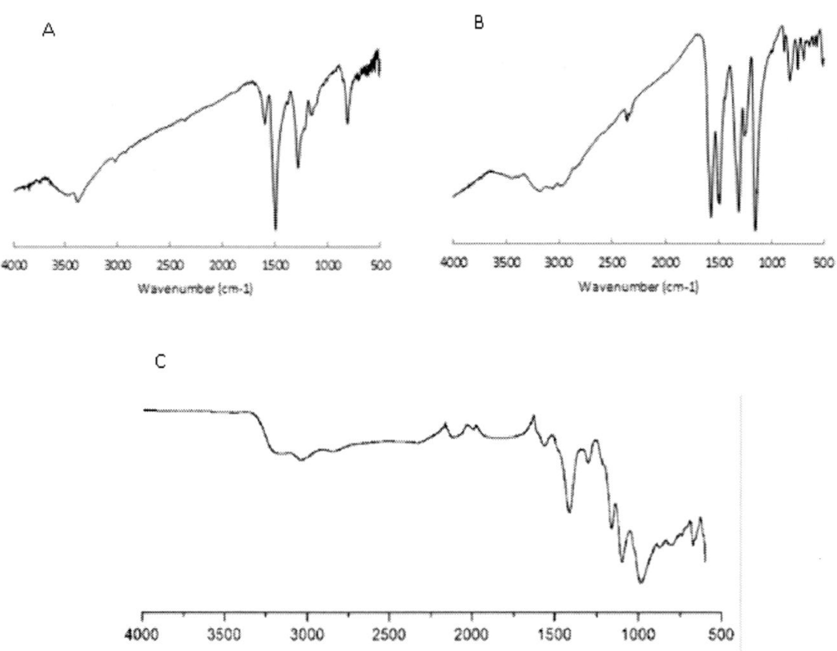

Figure 3.2. FT-IR (A), *emeraldine* (B) and *pernigraniline* (C).

The Fourier-Transform IR spectrum of emeraldine (Figure 3.2 B) shows a characteristic band at 1570 cm^{-1}, assigned to the C=C stretching of the quinoid rings (N=Q=N) and two peaks at 1498 cm^{-1} and 1484 cm^{-1}, assigned to the C=C stretching vibration mode for the benzenoid rings (N-B-N). The peaks at 1311 cm^{-1} and 1246 cm^{-1} are related to the C-N and C=N stretching modes and those at 1027 cm^{-1} and 889 cm^{-1} to the in-plane and out-of-plane bending of C-N. The peaks at 754 cm^{-1} and 692 cm^{-1} correspond to deformation vibration modes for the aromatic rings, while the peak at 573 cm^{-1} is characteristic for the 1, 4 di-substituted benzene. The ratio between the two bands at 1498 cm^{-1} and 1484 cm^{-1} is diagnostic to estimate the ratio between quinoid and aromatic groups and, consequently, the oxidation degree of the polymer. In fact, when polyaniline is in its totally reduced form (*leucoemeraldine*), this ratio is minor than one as *leucoemeraldine* is characterized by amino-benzenoid units. In its *emeraldine* form, polyaniline contains approximately the same amount of amino-benzenoid and imino-quinoid units and, as a consequence, the ratio is one (Figure 3.2. B). When polyaniline is in its totally oxidized form (*pernigraniline*) only imino-quinoid units are present in the backbone and the ratio is higher than 1.

The broad band from 2000 cm^{-1} to 4000 cm^{-1}, covering half the instrumental range, arises from the overlapping of many vibrational modes, especially the ones of NH and phenilene diamine groups. Moreover, some of the aminic nitrogen becomes protonated upon addition of an acid, thus becoming vastly hydrogen bonded and widening the band at high wavenumbers.

3.3. Atomic Absorption Spectroscopy (AAS)

Atomic Absorption Spectrometry (AAS) is an analytical technique that measures the concentrations of elements in a sample. Atomic absorption is so sensitive that it can measure down to parts per billion of a gram. The technique makes use of the wavelengths of light specifically absorbed by an element. They correspond to the energies needed to promote electrons from one energy level to another higher energy level. Atoms of different elements absorb characteristic wavelengths of light. Analysing a sample to see if it contains a particular element means using light from that element.

In AAS the sample is atomised, that is converted into ground state free atoms in the vapour state, and a beam of electromagnetic radiation emitted from excited element atoms is passed through the vaporised sample. Some of the radiation is absorbed by the element atoms in the sample. The greater the number of atoms there is in the vapour, the more radiation is absorbed proportionally to the number of the element atoms. A calibration curve is constructed by running a series of samples of known element concentration (standards) under the same conditions as the unknown. The amount the standard absorbs is compared with the calibration curve and this enables the calculation of the element concentration in the unknown sample. Atomic absorption spectrometry is successfully employed in different areas of chemistry (clinical and environmental analysis, pharmaceuticals, industry and mining) but it results particularly effective for detecting elements (the inorganic component) inside polymer composites.

3.4. Thermogravimetric Analysis (TGA)

Thermogravimetric analysis (TGA) is an analytical technique used to determine the thermal stability of a material and its fraction of volatile components by monitoring the weight change during a specimen heating. The measurement is normally carried out under air or inert atmosphere, such as helium or argon, and the weight is recorded as a function of increasing temperature. The maximum temperature is selected so that the specimen weight is stable at the end of the experiment, implying that all chemical reactions are completed. Sometimes, the measurement is performed in a lean oxygen atmosphere (1 to 5%O_2 in N_2 or He) to slow down oxidation. This technique can be effectively employed for assessing the thermal stability of polymer composites in comparison with the pristine polymeric material. As an example, in Figure 3.3. the thermogravimetric analyses of pristine polyaniline (sample 2) and some PANI/$CoFe_2O_4$ composites (samples 3, 4, 15) are reported (data unpublished).

The stabilizing effect of $CoFe_2O_4$ NPs is clearly detectable. It is known from the literature that cobalt is able to coordinate aminic nitrogen and thus $CoFe_2O_4$ interacts with the polymer chains. [153]

As a consequence, the stability of pure polyaniline (sample 2) is inferior if compared to its composites (sample 3 and 4, powder- and ferrofluid-form respectively) as its biggest step of decomposition takes place at lower temperatures. This is confirmed also in the case of the mechanical mixture between polyaniline and $CoFe_2O_4$ powder-type (sample 15). Here the contact between PANI and $CoFe_2O_4$ is not so intimate to allow the coordination, therefore the decomposition step takes place 100°C before the true composites (samples 3-4).

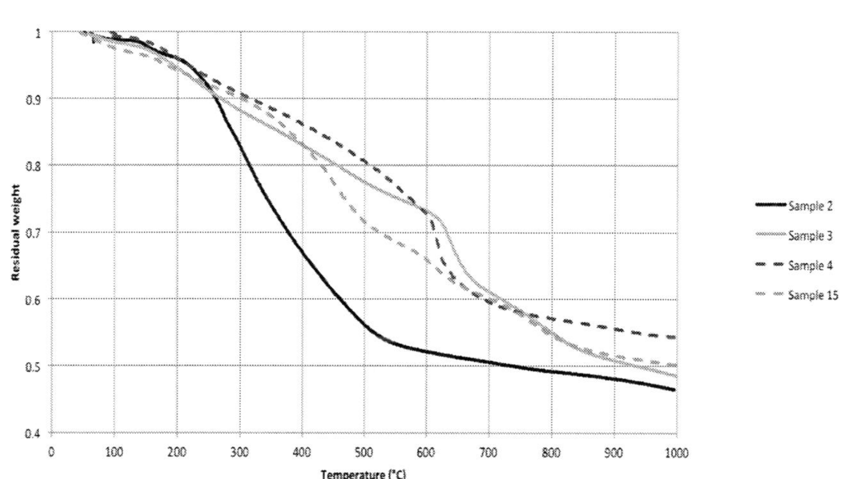

Figure 3.3. TGA of pure polyaniline and its $CoFe_2O_4$ composites.

3.5. Differential Scanning Calorimetry (DSC)

Differential Scanning Calorimetry, or DSC, is a thermal analysis technique that looks at how a material heat capacity (the amount of energy a unit of matter can hold) is changed by temperature. The difference in the amount of heat required to increase the temperature of a sample and reference is measured as a function of temperature. Both the sample and reference are maintained at nearly the same temperature throughout the experiment. Generally, the temperature program for a DSC analysis is designed such that the sample holder temperature increases linearly as a function of time. The reference sample should have a well-defined heat capacity over the range of temperatures to be scanned. A sample of known mass is heated or cooled and the changes in its heat capacity are tracked as changes in the heat flow. This allows the detection of transitions such as melts, glass transitions, phase changes and curing. Because of this flexibility, since most materials exhibit some sort of transitions, DSC is used in many industries, including pharmaceuticals, polymers, food, paper, printing, manufacturing, agriculture, semiconductors and electronics. The biggest advantage of DSC is the ease and speed with which transitions in materials can be seen. In liquid crystals, metals, pharmaceuticals and pure organics, it is possible to see phase changes or polymorphs and study the degree of purity in materials. While processing or distilling materials, knowledge of a material heat capacity and heat content change (called enthalpy) can be used to estimate how efficiently the process is operating. This technique is particularly powerful for determining the glass transition of polymeric materials, that is the reversible transition in amorphous materials or in amorphous areas within semicrystalline materials from a hard and brittle state into a molten or rubber-like state.

3.6. Scanning Electron Microscopy (SEM)

The scanning electron microscopy (SEM) produces images of a sample by scanning it with a focused beam of electrons. The high-energy electrons interact with atoms in the sample to generate a variety of signals at the surface of solid specimens, thus revealing information about external morphology (texture), chemical composition, crystalline structure and orientation of materials making up the sample. A 2-dimensional image is generated that displays spatial variations in these properties by collecting data over a selected area of the surface. Areas ranging from 1 cm to 5 microns in width can be imaged in a scanning mode using conventional SEM techniques, allowing magnification ranging from 20X to approximately 30000X. The SEM is also able to analyze selected point locations on the sample. This approach is particularly useful in qualitatively or semi-quantitatively determining chemical compositions (using EDS).

3.7. Transition Electron Microscopy (TEM)

The transmission electron microscope (TEM) operates on the same basic principles as the light microscope but uses electrons instead of light. What one can see with a light microscope is limited by the wavelength of light. Conversely, TEM uses electrons as "light source" and their much lower wavelength makes it possible to get a resolution a thousand times better than with a light microscope. The electrons beam is transmitted through an ultra-thin specimen, interacting with the specimen as it passes through. An image is formed from the interaction of the electrons transmitted through the specimen, the image is magnified and focused onto an imaging device, such as a fluorescent screen, on a layer of photographic film. This enables to examine fine detail, even as small as a single column of atoms, which is thousands of times smaller than the smallest resolvable object in a light microscope. TEM represents a major analysis method in a range of scientific fields, in both physical and biological sciences. TEM finds application in cancer research, virology, materials science as well as pollution, nanotechnology and semiconductor research.

3.8. X-Ray Powder Diffraction (XRPD)

X-Ray Powder Diffraction (XRPD) is one of the most rapid and powerful analyses used to investigate the structure of a solid in terms of lattice constants, phase identification of a crystalline material as well as unit cell dimensions. The analyzed material is finely ground, homogenized and average bulk composition is determined. X-ray diffraction is based on constructive interference of monochromatic X-rays and a crystalline sample, which occurs when the interaction of the incident rays with the sample satisfy the Bragg's Law ($n\lambda = 2d \sin \theta$). This law relates the wavelength of electromagnetic radiation (λ) to the diffraction angle (θ) and the lattice spacing (d) in a crystalline sample. Such diffracted X-rays are then detected, processed and counted. All possible diffraction directions of the lattice can be attained by scanning the sample through a wide range of 2θ angles, due to the random orientation of the powdered material. Since each mineral has a set of specific d-spacings, converting the diffraction peaks to d-spacings allows identification of the material. Typically, this is achieved by comparison of d-spacings with standard reference patterns.

This technique can be successfully employed for disentangle also polyaniline structure and its composites. X-Ray diffraction, in fact, enables to investigate the chain ordering and crystallinity in polymeric materials. As mentioned in Section 1.4, a high level of conductivity in PANI can be obtained by doping the polymer with aqueous HCl. Moreover, each conducting polymer particle can be considered as a conducting crystal grain whose size and degree of crystallinity dramatically affect the conductivity (Section 1.5). In general, no distinctive crystalline structure is observed in undoped PANI. In this case an amorphous peak appears at $2\theta \sim 20°$. Studying the morphology of conducting polymer, it was found that the ratio of half-width to height (HW/H) of the X-ray diffraction peak reflects ordering in the polymer backbone. Accordingly, small HW/H values correspond to high crystalline order. After doping, a small sharp peak appears at $2\theta \sim 9.0°$, which can be attributed to the crystallinity, while the peak at $2\theta \sim 20°$ is shifted to higher angle, corresponding to a decreased d-spacing between polymer backbones. This indicates that the doping process leads to a more compact chain structure while enhancing PANI crystallinity.

3.9. Conductivity Measurement

The electrical resistivity (or its reciprocal, conductivity) of a material can be measured by means of simple resistance probe. Both two-probe and four-probe methods are the most commonly variants employed for the scope. In the former method, a uniform current density is applied across the specimen sandwiched between two electrodes located on parallel faces, the potential drop across the electrodes is then measured. According to this scheme, current injection and voltage measurement are made at the same two electrodes and the resistivity can be measured when the material specimen has a simple geometry of known dimensions. In the four-probe method, one pair of probes is employed for the current injection at a pair of electrodes whereas the other pair is used for voltage measurements on another set of electrodes. The resistivity can be calculated if the applied current is uniform and the specimen dimensions are known. Comparing the two methods, the four-probe is not sensitive to contact resistance issues thereby resulting in a more reliable tool for conductivity measurements.

The conductivity of polymers is usually measured as resistivity on compressed powder or film samples. The theoretical principle is the *Kelvin bridge*. Basically, the two outer probes force a current passing into the sample and the inner ones measure a potential. Accordingly, resistivity (ρ) can be obtained using the following equations (Eq. 3.1.- 3.2.).

$$\rho\ (\Omega\ \mathrm{cm}) = (2\ \pi\ s\ V)\ /\ I \qquad \text{(Eq. 3.1.)}$$

where s = distance between the sensing points, V = potential, I = current.

In particular, this can be used when the instrument has the four probes equally spaced and the thickness t is much bigger than the distance between the sensing points s.

Conversely, when thickness t is much smaller than the distance between the sensing points s, i.e., in films, Eq. 3.1. changes into Eq. 3.2.

$$\rho\ (\Omega\ \text{cm}) = (\pi\ t\ V)\ /\ \ln 2I \qquad \text{(Eq. 3.2.)}$$

3.10. Cyclic Voltammetry (CV)

Voltammetry is one of the techniques which electrochemists employ to assess electrolysis mechanisms, thus attaining information about an analyte by measuring the current as the potential is varied. The potential (V) is changed arbitrarily either step by step or continuously and the correspondent current value (I) is measured as the dependent variable, thereby originating curves I = f(V) called voltammograms. To carry out such an experiment, an electrolysis cell needs to be assembled. This requires at least two electrodes: the *working electrode*, allowing the contact with the analyte and the desired potential to be applied, and a *reference electrode* acting as the other half of the cell. This latter electrode must have a known potential with which to gauge the potential of the working electrode. Both are hooked up to an external electrical circuit, even though it is only the working electrode that controls the flow of current in the electrochemical measurement.

There are numerous forms of voltammetry:

- Potential Step Voltammetry
- Linear Sweep Voltammetry
- Cyclic Voltammetry

For each of these variants, a voltage or series of voltages is applied to the electrode and the actual current that flows monitored but CV is the most widely used technique as it offers a rapid evaluation of redox potentials of the electroactive species. In a typical CV experiment, the working electrode potential is swept linearly versus time between two values (i.e., V1 and V2) at a fixed rate like the linear sweep voltammetry (LSV). However, differently from LSV, when the voltage reaches the set value V2 the scan is reversed and the voltage is swept back to V1. This inversion can happen multiple times during a single experiment. When the scan is reversed the electrolysis product is gradually converted to the reactant through the equilibrium positions. This means the current flow is now from the solution species back to the electrode.

The current generated at the working electrode is then plotted versus the applied voltage to give the cyclic voltammogram.

As reported by Snauwaert et al. the ratio between the amine and imine content in polyaniline is function of the electrochemical potential. [154] Hence, CV appears to be particularly indicated for detecting the different oxidation states leading to the three most common forms of PANI (*leucoemeraldine, emeraldine* and *pernigraniline*).

Chapter 4

APPLICATIONS OF POLYANILINE AND ITS MAIN COMPOSITES: FROM TRADITION TO INNOVATION

The versatile properties of polyaniline can open the way to a wide spectrum of applications, ranging from diodes, transistors and sensors to drug delivery systems. However, to date, the valuable scientific achievements reported in the literature have not yet found adequate application in industry.

A notable progress has been reached in two specific fields, namely EMI (electromagnetic interference) shielding and force sensors which will be briefly presented in the following sections.

4.1. ELECTROMAGNETIC PANI COMPOSITES FOR EMI SHIELDING

Owing to the huge number of electronic devices sold every year and used daily, in the last years electromagnetic interference has become a serious concern for health, causing insomnia, nervousness and headache, but also for the electronic signals transmission.

EMI shielding involves three main mechanisms: reflection, absorption and multiple-reflection. [155] Metal-based shields are particularly effective in interacting with the incoming EM waves thanks to their mobile charge carriers (electrons or holes). [156] However, the weight and stiffness of such materials limit their application in aerospace industry, tissues and so on. A satisfactory alternative is represented by organic-inorganic composites since they combine the advantages of inorganic materials (i.e., electrical and magnetic properties, thermal and mechanical stability) with those of organic polymers (i.e., lightness, flexibility, processability) making them good candidates for EMI shielding. The ideal material able to interact with EM waves should have both electrical or magnetic dipoles. The electro-conductive component shields the electromagnetic waves by absorption mechanism based on resistive loss which consists in transforming the electromagnetic energy into heat by Joule effect. Due to their high permeability and ability to absorb microwaves, even magnetic materials (i.e., ferrites) contribute to EMI shielding of electrical and magnetic composites. EMI shielding in polymer composites is more complicated than in homogeneous materials, because of the wide surface area available for reflection and multiple reflection, the latter being the re-reflection of waves already reflected (Figure 4.1). [156]

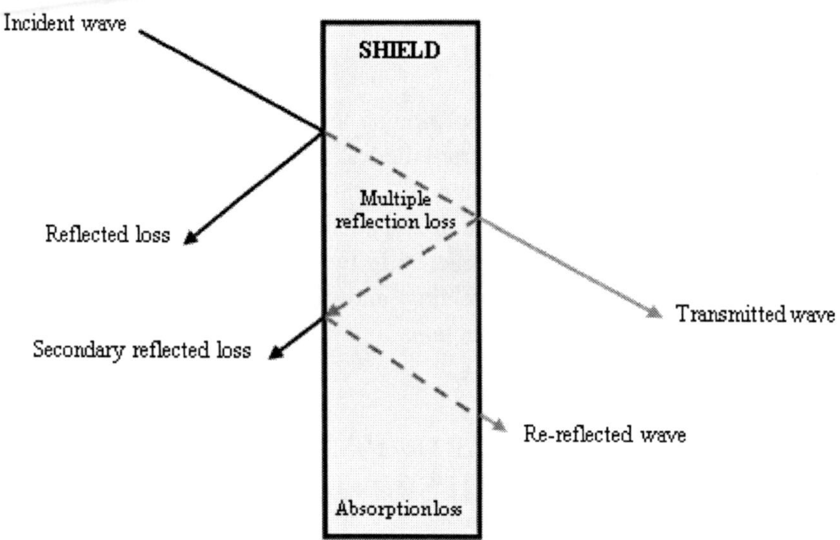

Figure 4.1. Basic shielding mechanism.

SE (shielding effectiveness) in decibel (dB) is a measure of EMI reduction at a specific frequency achieved by a shield and it is defined according to Equation 4.1:

$$SE = 10\log\frac{Po}{Pt} = 20\log\frac{Eo}{Et} = 20\log\frac{Ho}{Ht} \quad \text{(Eq. 4.1)}$$

where Po, Eo and Ho are respectively power, electrical and magnetic fields of the incident wave, while Pt, Et and Ht are the counterparts transmitted through the shield. [157]

Among the materials used for this scope, PANI and PANI composites, consisting of magnetic fillers embedded in the polymeric matrix, are intensively investigated.

In the case of pristine PANI, many parameters can affect its performances. Accordingly, SE increases linearly with thickness when employing a doped PANI sample in form of film with thickness higher than 10 μm. On the contrary, if film is thin (<10 μm) SE variation is not linear, probably because of numerous inner multiple reflections. Moreover, temperature has a negligible effect on PANI performances, whereas SE is improved by polymer stretching. Differently from metals, whose unique important parameter influencing SE is conductivity, for PANI and, in general, for all the ICPs also permittivity and tan δ (which is the ratio of the imaginary part to the real part of the dielectric constant) play a key role. In fact, it has been demonstrated that large values of tan δ and permittivity make PANI an excellent absorber of electromagnetic radiations. [158] Recently, Tantawy and coworkers compared the electromagnetic shielding properties of three PANI samples prepared according to three different approaches: fast and slow oxidation by a traditional synthesis and an innovative solvent-free method. [159] They found how the synthetic approach can markedly affect EM shielding effectiveness, in particular in the X-band. Hence, PANI samples prepared following the solvent-free or limited solvent approach exhibited crystallinity levels, conductivity values and SE higher than those displayed by PANI samples synthesized by the traditional protocols.

PANI/nanomagnets composites have received a great deal of attention thanks to their peculiar characteristics. They can be prepared following several protocols, mainly consisting in the *ex situ* methods involving multiple reaction steps. Typically, magnetic nanomaterials (generally in form of nanoparticles) are first synthesized, then PANI is separately prepared by oxidative polymerization of aniline monomer and, finally, all the compounds are mixed together. [160]

Although this is the most common and easiest method to prepare this kind of materials, many other syntheses have been developed, as reported below. Omissions are inevitable due to the myriad of investigations carried out on this subject. The major challenge in PANI/nanomagnets preparation is the control of nanoparticles dispersion inside the polymer matrix, since particle–particle and particle–matrix interactions often lead to clustering of NPs. In order to reach a good dispersion of NPs, these latter can be prepared *in situ* [161] or used under specific conditions in form of ferrofluid. [120] However, in both cases, high speed of stirring and proper surfactants are required in order to guarantee small size and low aggregation level of inorganic NPs.

Electrical and magnetic PANI/Fe_xO_y composites can be easily prepared by a chemical method, mixing EB in NMP with aqueous iron sulfate solutions at room temperature under nitrogen atmosphere at different pH. [121] In their work the authors demonstrated that Fe_xO_y loading in the composites can be tuned choosing the right pH values (high Fe_xO_y loadings were observed for high pH).

Among the nanomaterials employed as magnetic fillers for the preparation of PANI/nano-magnets composites, nano-ferrites are the most used and effective for EMI shielding.

The absorption properties of PANI/$CoFe_2O_4$ nanocomposites were measured in the range 12.4-18 GHz in Ku-band and reached SE higher than 21.5 dB, corresponding to more than 99% microwaves attenuation. [162] PANI matrices embedding $MnFe_2O_4$ [163] and Fe_3O_4 [164] were tested at different frequency ranges and in all cases the electromagnetic composites exhibited more promising results than the pristine conducting polymer.

Although several ecofriendly routes are now available for preparing PANI/nanomagnets composites, no results as microwave absorbers have been yet reported in the literature.

4.2. Stress/strain Sensors

Thanks to their efficiency and sensitivity, stress/strain sensors have found application in the manufacturing of materials for air conditioning, heating and ventilation systems, petrochemical plants, but also innovative touch technology. Moreover, they are particularly important for monitoring tire pressure, engine management and automatic transmission, as well as for medical applications (e. g., infusion pumps, intracranial measurement) and many devices widely used in everyday life and leisure (e. g., scuba gears, appliances). For their versatility, high performances and low cost the market of stress/strain sensors is continuously growing and the forecasts are of about $7.34 billion in 2017. Even though metal-based traditional materials are still employed in this sector, they fail for many aspects: low flexibility, heaviness, delamination and so on. Hence, they cannot be employed in fields where flexibility, stretching and lightness along with high sensitivity are dominant (e.

g., robotic, biomechanics). Recently, conducting organic polymers have gradually become a good alternative. The main characteristic making them ideal materials is the piezoresistivity, that is the ability of a material to modify its electrical resistivity when a stress or strain is applied. The magnitude of piezoresistive effect is quantified by a parameter called gauge factor (GF) as expressed by Equation 4.2:

$$GF = \frac{\Delta R/R0}{\varepsilon} \qquad \text{(Eq. 4.2)}$$

where ΔR is the change in resistance, R_0 is the steady-state resistance and ε the deformation.

Both extrinsically (ECPs) and intrinsically conducting polymers (ICPs) have been investigated as basic materials for piezoresistors. ECPs are based on conductive fillers embedded in an insulating polymer matrix. Although carbon nanofillers, including carbon black, nanofibers, nanotubes or graphene, are the most employed, even PANI fillers have been proposed in order to overcome the main drawback associated with PANI intractability. [165]

In these materials electrical conductivity is exclusively related to the presence of conducting nanofiller which, however, has an effect on the mechanical, electrical and thermal properties of the resulting composite. Near the electrical percolation threshold (the concentration of the conducting filler after which the change in resistivity becomes marginal), electroconductivity reaches its maximum value as well as the GF, depending on polymer matrix but also on type and amount of nanofiller employed.

In this context, interesting results were obtained by Rahaman and coworkers for PANI/EVA (ethylene vinyl acetate) composites under compression (up to 73 kPa). [166] As expected, the electrical resistivity decreased by increasing load applied and PANI content and percolation threshold was reached for PANI concentration of 30-35% parts of PANI by weight per hundred parts of EVA. Concerning the pressure effect, the authors highlighted that only at the beginning (0-35 kPa) a strong decrease in resistivity was observed whereas, by gradually increasing the applied load, a change in resistivity was marginal. Such an effect might be due to the reduced thickness/volume ratio of the samples caused by pressure, thus leading to more compact structures. During the compression processes, in fact, the inter-particle distance among the conducting meshes is gradually reduced leading to a new conducting network inside the composites. However, at the same time, an excessive load can destroy some of the already existing conducting networks. The authors concluded that at low loads (up to 35 kPa) the first contribution is dominant and positively affects the conduction mechanism in the samples but when the applied loads are too high (higher than 35 kN) both the contributions become equivalent and the resistivity remains almost unchanged. In terms of sensitivity, the best results were obtained for composites with low PANI content, probably because for such samples the number of already existing conducting networks is less and other ones are obtained under compression. As a result, the application of pressure produces high change in resistivity. By progressively increasing PANI content, the number of already existing conductive networks in the polymer matrix grows up and, hence, the formation of new conductive meshes during the loading process is more modest, thereby causing a strong reduction in the resistivity change. The equilibrium among existing and subsequently formed conducting networks seems also responsible for the effect of time on resistivity under constant pressure. The resistivity strongly and rapidly decreases at the beginning of the analyses. However, after some time the change is marginal, finally reaching an asymptotic behavior.

In addition to ethylene vinyl acetate (EVA) many other polymers have been investigated as insulating matrix. Among them, poly(vinyl acetate) [167] and poly(vinylidene)fluoride (PVDF). [165]

The piezoresistive properties of polyaniline, but in general of all ICPs, are related to its conduction mechanism consisting of a double charge movement: along the polymer backbone (*intra*-chain mechanism) and between different polymer chains (*inter*-chain mechanism) (see Section 1.6).

This second contribution becomes particularly important when the material is subjected to a force or a pressure and is responsible for its piezoresistivity. Pristine polyaniline in form of pellets and films have been extensively investigated by many authors under different conditions and for these reasons a comparison of the results is not possible.

In any case, a short view of the most interesting achievements is reported below.

A complete and accurate study on the effect of pressure and pressing time on PANI pellets was recently realized by Zhang. [168] Using *p*-toluensulfonic acid as the dopant, Zhang reported that increasing the applied pressure from 35 to 70 MPa, electroconductivity of PANI pellets gradually increased reaching the maximum value of 7.56 S/cm. Moreover, conductivity of all the pellets grew up with the pressing time, because this latter has an effect on the removal of the voids inside the pellets, thereby leading to a more and more compact structure. However, when the applied pressure exceeded the critical value of

70 MPa, a slight conductivity decrease was observed, probably due to a modification of polymer crystallinity.

Recently, a comparison between fresh and thermally aged polyanilines prepared by two different methods was proposed by Della Pina et al.: a traditional method, based on aniline oxidative polymerization by amoniumperoxide, and a green route regarding the oxidative polymerization of *N*-(4-aminopheyl)aniline by hydrogen peroxide. [169] In particular, they investigated the conductivity variation of both the materials after thermal aging in a temperature range from 25 to 200°C under compression/expansion processes in the range 0-750 MPa. Also after thermal treatment, the traditional PANI-based pellets ritained their high conductivity and low hysteresis although characterized by low sensitivity. On the contrary, the electroconductivity of "green" PANI pellets was strongly dipendent on thermal aging. In fact, beside showing big hysteresis, low values of conductivity and high pressure sensitivity, these materials exhibited a drastic drop of conductivity while temperature increased. Their different behaviour seems to be related to water content inside the samples. Actually, TGA analyses showed higher mass loss after T = 100°C for "green" PANI pellets with respect to those prepared by the traditional protocol, therefore justifying a major thermal instability.However, further and more accurate investigations emphasized the higher electromechanical response for "green" PANI than for the traditional one, thus encouraging potential applications of this innovative material as sensing device for force or pressure measuring systems. [170] The authors demonstrated the electromechanical behaviour of both kinds of polyaniline to be deeply dependent on their crystallinity degree. In particular, the high values of conductivity as well as high brittleness of the traditional PANI are attributable to its semi-crystallinity having a negative effect on its mechanical performances. On the contrary, even though the amorphous "green" PANI is characterized by low values of conductivity, its higher mechanical performance makes it more suitable for pressure and force measurements. The low GF values observed for both types of PANI were attributed to two main factors: pellet form and small dimension of the doping agent employed (H_2SO_4). In fact, both these factors are able to reduce the contribution of the*inter*-chain mechanism to the whole charges movement inside the materials, in the first case because of the random orientation of the polymer chains in each pellet, in the second case because of the small size of the dopant which reduces the *inter*-chain displacement.

Conclusion

Polyaniline and its composites, the most popular conducting materials whose unique properties and applications are drawing a worldwide interest, have been the focus of this book. Their main synthetic methods have been presented, ranging from the traditional protocols to the innovative eco-friendly routes which employ "green" oxidants and proper catalysts. A brand new preparation protocol able to facilitate the polymer processing into pure PANI electrospun nanofibers has even been described. The possibility to combine the fascinating conducting properties of polyaniline with those of other materials has opened the way to the preparation of amazing composite materials. PANI nanocomposites and blends have found a particular place inside this book, together with the principal analytical techniques for characterizing such materials and an updated description of their huge spectrum of applications. Polyaniline and its composites, in fact, are already successfully employed in many electrical and optical devices, as well as advanced gas sensors, but novel emerging usages in EMI shielding and force sensors are now showing their great potentiality. Academics, researchers, scientists, engineers and students in the field of material science, chemistry, physics, electronics and nanotechnology will benefit from this book which reports the updated accomplishments in polyaniline research and is application-oriented.

REFERENCES

[1] Runge, F. F. Ueber einige Produkte der Steinkohlendestillation. *Annalen der Physik und Chemie* 1834, *31*, 513-524.

[2] Meth-Cohen, O.; Smith, M. What did W. H. Perkin actually make when he oxidised aniline to obtain mauveine? *J. Chem. Soc., Perkin Trans.*1994, *1*, 5-7.

[3] Seixas de Melo, J.; Takato, S.; Sousa, M.; Melo, M. J.; Parola, A. J. Revisiting Perkin's dye(s): the spectroscopy and photophysics of two new mauveine compounds (B2 and C). *Chem. Commun.* 2007, 2624-2626.

[4] Heichert, C.; Hartmann, H. On the formation of mauvein: mechanistic monsiderations and preparative results. *Z. Naturforsch*, 2009, *64b*, 747-755.

[5] MacDiarmid, A.G.; Epstein, A. J. Polyanilines: a novel class of conducting polymers. *Faraday Discuss. Chem. Soc.*1989, *88*, 317-332.

[6] Green, A.; Woodhead, A. CCXLIII.—Aniline-black and allied compounds. Part I. *J. Chem. Soc., Trans.* 1910, *97*, 2388-2403.

[7] Green, A.; Woodhead, A. CXVII.—Aniline-black and allied compounds. Part II *J. Chem. Soc., Trans.* 1912, *101*, 1117-1123.

[8] Huang, J.; Kaner, R. B. A general chemical route to polyaniline nanofibers. *J. Am. Chem. Soc.* 2004, *126*, 851-855.

[9] Beau, B.; Travers, J. P.; Banka, E. NMR evidence for heterogeneous disorder and quasi-1D metallic state in polyaniline CSA. *Synth. Met.*1999, *101*, 772-775.

[10] Papathanassiou, A. N.; Sakellis, I.; Grammatikakis, J.; Sakkopoulos, S.; Vitoratosb, E.; Dalas, E. An insight into the localization of charge

carriers in conducting polyaniline by analyzing thermally stimulated depolarization signals. *Solid State Commun.* 2003, *125*, 95-98.

[11] Wnek, G. A proposal for the mechanism of conduction in polyaniline. *Synth. Met.* 1986, *16*, 213-218.

[12] Catedral, M. D.; Tapia, A.K.G.; Sarmago, R. V.; Tamayo, J. P.; del Rosario, E. Effect of dopant ions on the electrical conductivity and microstructure of polyaniline (emeraldine salt). *Science Diliman,* 2004, *16*, 41-46.

[13] Zhou, Q.; Wang, J.; Ma, Y.; Cong, C.; Wang, F. The relationship of conductivity to the morphology and crystallinity of polyaniline controlled by water content via reverse microemulsion. *Colloid Polym. Sci.* 2007, *285*, 405-411.

[14] Angelopoulos, A; Ray, A; MacDiarmid, G. Polyaniline: processability from aqueous solutions and effect of water vapor on conductivity. *Synth. Met.*1987, *21*, 21-30.

[15] Travers, J. P.; Nechtschein, M. Water effects in polyaniline: A new conduction process. *Synth. Met.* 1987, 21, 135-141.

[16] Ohtani, A; Abe, M.; Ezoe, M.; Doi, T.; Miyata, T.; Miyake, A. Synthesis and properties of high-molecular-weight soluble polyaniline and its application to the 4MB-capacity barium ferrite floppy disk's antistatic coating. *Synth. Met.* 1993, *57*, 3696-3701.

[17] Camalet, J.; Lacroix, J.; Ngyen, T.; Aeiyach, S.; Chane-Ching, K.; Petit, J. J.; Chauveau, E.; Lacaze, P. Aniline electropolymerization on mild steel and zinc in a two-step process. *J. Electroanal. Chem.* 2000, *481*, 76-81.

[18] Stejskal, J.; Sapurina, I.; Trchová, M. Polyaniline nanostructures and the role of aniline oligomers in their formation. *Prog. Polym.Sci.* 2010, *35*, 1420-1481.

[19] Mandić, Z.; Duić, L.; Kovaćićek, F. The influence of counter-ions on nucleation and growth of electrochemically synthesized polyaniline film. *Electrochim. Acta* 1997, *42*, 1389-1402.

[20] Zotti, G.; Comisso. N. Electrodeposition of polythiophene, polypyrrole and polyaniline by the cyclic potential sweep method. *J. Electroanal.Chem. Interfacial Electrochem.* 1987, *235*, 259-273.

[21] Mu, S.; Chen, C.; Wang, J. The kinetic behavior for the electrochemical polymerization of aniline in aqueous solution. *Synth. Met.* 1997, *88*, 249-254.

[22] Park, S.; Joong, H. Recent Advances in Electrochemical Studies of π-Conjugated Polymers. *Bull. Korean Chem. Soc.* 2005, *26*, 697-706.

[23] Planes, G. A.; Rodríguez, J. L.; Miras, M. C.; García, G.; Pastore, E.; Barber, C. A. Spectroscopic evidence for intermediate species formed during aniline polymerization and polyaniline degradation. *Phys. Chem. Chem. Phys.* 2010, *12*,10584-10593.

[24] Geniès, E. M.; Boyle, A.; Lapkowski, M.; Tsintavis, C. Polyaniline: A historical survey. *Synth. Met.* 1990, *36*, 139-182.

[25] Mohilner, D. M.; Adams, R. N.; Argersinger, W. J. Investigation of kinetics and mechanism of anodic oxidation of aniline in aqueous sulfuric acid solution at a platinum electrode. *J. Amer. Chem. Soc.* 1962, *84*, 3618-3622.

[26] Bacon, J.; Adams, R. N. Anodic oxidations of aromatic amines. III. Substituted anilines in aqueous media. *J. Amer. Chem. Soc.* 1968, *90*, 6596-6599.

[27] Oyama, N.; Ohsoka. T. Electrochemical properties of the polymer films prepared by electrochemical polymerization of aromatic compounds with amino groups. *Synth. Met.* 1987, *18*, 375-380.

[28] Yu, L. T.; Borredon, M. S.; Jozefowicz, M.; Belorgey, G.; Buvet, R. Experimental Study of direct current conductivity of macromolecular compounds. *J. Polym. Sci. Part C-Polym. Symp.* 1967, *16*, 2931.

[29] Sarma, T. K.; Chowdhury, D.; Paul, A.; Chattopadhyay, A. Synthesis of Au nanoparticle-conductive polyaniline composite using H_2O_2 as oxidising as well as reducing agent. *Chem. Commun.* 2002, 1048-1049.

[30] Bicak N., Karagoz, B. Polymerization of aniline by copper-catalyzed air oxidation. *J. Polym.Sci.: Part A: Polymer Chemistry*. 2006, *44*, 6025–6031.

[31] Sun, Z.; Geng, Y.; Li, J.; Jing, X.; Wang, F. Chemical polymerization of aniline with hydrogen peroxide as oxidant. *Synth. Met.* 1997, *84*, 99-100.

[32] MacDiarmid, A. G. ; Chiang, J. C.; Halpern, M.; Huang, W. S.; Mu, S. L.; Somasiri, N. L. D; Wu, W.; Yaniger , S. I. “Polyaniline”: Interconversion of metallic and insulating forms *Mol. Cryst., Liq. Cryst* 1985, *121*, 173-180.

[33] Cao, Y.; Andreatta, A.; Heeger, A. J.; Smith, P. Influence of chemical polymerization conditions on the properties of polyaniline. *Polymer* 1989, *30*, 2305-2311.

[34] Österholm, J. E.; Cao, Y.; Klavetter, F.; Smith, P. Emulsion polymerization of aniline. *Synth. Met.* 1993, *35*,2902–2906.

[35] Kinlen P. J.; Liu, J.; Ding, Y.; Graham, C. R.; Remsen, E. E. Emulsion polymerization process for organically soluble and electrically conducting polyaniline. *Macromolecules* 1998, *31*,1735–1744.

[36] Kinlen ,P. J.; Frushour, B. G.; Ding, Y.; Menon, V. Synthesis and characterization of organically soluble polyaniline and polyaniline block copolymers. *Synth Met.* 1999, *10*,758–761.

[37] Paul, R. K.; Veena, V.; Pillai C. K. S. Melt/solution processable conducting polyaniline: doping studies with a novel phosphoric acid ester. *Synth Met.* 1999,*104*,189–195.

[38] Paul R. K.; Pillai C. K. S. Thermal properties of processable polyaniline with novel sulfonic acid dopants. *Polym. Int.* 2001, *50*, 381–386.

[39] Wei Z.; Wan M. Hollow microspheres of polyaniline synthesized with an aniline emulsion template. *Adv Mater.* 2002, *14*, 1314–1317.

[40] Kim, B. –J.; Oh, S. –G.; Han, M. –G.; Im, S. -S. Preparation of polyaniline nanoparticles in micellar solutions as polymerization medium. *Langmuir* 2000, *16*, 5841–5845.

[41] Swapna Rao, P.; Sathyanarayana, D. N.; Palaniappan S. Polymerization of aniline in an organic peroxide system by the inverse emulsion process. *Macromolecules* 2002, *35*, 4988–4996.

[42] Sai Ram, M.; Palaniappan, S. A process for the preparation of polyaniline salt doped with acid and surfactant groups using benzoyl peroxide. *J Mater Sci* 2004, *39*, 3069–3077.

[43] Shreepathi, S; Holze, R. Spectroelectrochemical investigations of soluble polyaniline synthesized via new inverse emulsion pathway. *Chem Mater.* 2005, *17*, 4078–4085.

[44] Marie, E.; Rothe, R.; Antonietti, M.; Landfester, K. Synthesis of polyaniline particles via inverse and direct miniemulsion. *Macromolecules* 2003, *36*, 3967–3973.

[45] Li, J.; Fang, K.; Qiu, H.; Li, S.; Mao, W.; Wu, Q. Micromorphology and conductive property of the pellets prepared by HCl-doped polyaniline nanofibers. *Synth Met.* 2004, 145, 191-194.

[46] Kuo, C. -W.; Wen, T. -C. Dispersible polyaniline nanoparticles in aqueous poly(styrenesulfonic acid) via the interfacial polymerization route. *Eur. Polym. J.*, 2008, *44*, 3393-3401.

[47] Likhar, P. R.; Arundhathi, R.; Ghosh, S.; Lakhmi Kantam, M. Polyaniline nanofiber supported $FeCl_3$: An efficient and reusable heterogeneous catalyst for the acylation of alcohols and amines with acetic acid. *J. Mol. Catal. A Chem.* 2009, 3*02*, 142-149.

[48] Yan, Y.; Yu, Z.; Huang, Y. W.; Yuan, W. X.; Wei, Z. X. Helical Polyaniline Nanofibers Induced by Chiral Dopants by a Polymerization Process. *Adv. Mater.* 2007, *19*, 3353-3357.

[49] Huang, J. X.; Kaner, R. B. Nanofiber Formation in the Chemical Polymerization of Aniline: A Mechanistic Study. *Angew. Chem. Int. Ed.* 2004, *43*, 5817–5821.

[50] Guo, Q.; Yi, C.; Zhu, L.; Yang, Q.; Xie, Y. Chemical synthesis of cross-linked polyaniline by a novel solvothermal metathesis reaction of p-dichlorobenzene with sodium amide. *Polymer*, 2005, *46*, 3185-3189.

[51] Mohammadi, A., Hasan, M. -A.; Liedberg, B.; Lundström, I.; Salaneck, W. R. Chemical vapour deposition (CVD) of conducting polymers: polypyrrole. *Synth. Met.* 1986, *14*,189-197.

[52] Bhrada, S.; Lee, J. H. Synthesis of higher soluble nanostructured polyaniline by vapor-phase polymerization and determination of its crystal structure. *J. Appl. Polym. Sci.* 2009, *114*, 331-340.

[53] Suslick, K. S. Sonochemistry. *Science* 1990, *247*, 1439–1445.

[54] Jing, X.L.; Wang, Y.Y.; Wu, D; She, L.; Guo, Y. Polyaniline nanofibers prepared with ultrasonic irradiation. *J. Polym. Sci. Polym. Chem.* 2006, *44*, 1014–1019.

[55] Wang, Y.; Jing, X.; Kong, J. Polyaniline nanofibers prepared with hydrogen peroxide as oxidant. *Synth. Met.* 2007, *157*, 269–275.

[56] Toshima, N.; Yan, H.; Ischiwatari, M. Catalytic Polymerization of Aniline and Its Derivatives by Using Copper(II) Salts and Oxygen. New Type of Polyaniline with Branched Structure. *Bull. Chem. Soc. Jpn.* 1994, *67*, 1947.

[57] Dias, H. V. R.; Wang, X.; Rajapakse, R. M. G.; Elsenbaumer, R. L. A mild, copper catalyzed route to conducting polyaniline. *Chem. Commun.* 2006, 976-978.

[58] Chen, Z.; Della Pina, C.; Falletta, E.; Rossi, M. A green route to conducting polyaniline by copper catalysis. *J. Catal.* 2009, *267*, 93-96.

[59] Sadighi, J. P.; Singer, R. A.; Buchwald, S. L. *J. Amer. Chem. Soc.* 1998, *120*, 4960-4976.

[60] Nadagouda, M. N.; Varma, R. S. Green synthesis of Ag and Pd nanospheres, nanowires and nanorods using vitamin B2: catalytic polymerization of aniline and pyrrole. *J. Nanomater.* 2008, 1-8.

[61] Mallick, K.; Witcomb, M. J.; Scurrell, M. S. Gold in polyaniline: Recent trends. *Gold Bull.* 2006, *39*, 166-174.

[62] Chen, Z.; Della Pina, C.; Falletta, E.; Lo Faro, M.; Pasta, M.; Rossi, M.; Santo, N. Facile synthesis of polyaniline using gold catalyst. *J. Catal.* 2008, *259*,1-4.
[63] de Barros, R. A.; de Azevedo, W. M.; de Aguiar, F. M. Photo-induced polymerization of polyaniline. *Mater. Charact.* 2003, *50*, 131-134.
[64] Khanna, P. K.; Sigh, N.; Charan, S.; Viswanath, A. K. Synthesis of Ag/polyaniline nanocomposite via an in situ photo-redox mechanism. Mater. Chem. Phys. 2005, *92*, 214-219.
[65] Bourdo, S. E.; Berry, B. C.; Viswanathan, T. Catalytic effects of selected transition metal ions in the synthesis of lignosulfonic acid doped polyaniline. J. Appl. Polym. Sci. 2005, *98*, 29-33.
[66] Fong, Y.; Schlenoff, J. B. Polymerization of aniline using mixed oxidizers. *Polymer*, 1995, *36*, 639-643.
[67] Tzou, K.; Gregory, R. V. Kinetic study of the chemical polymerization of aniline in aqueous solutions. *Synth. Met.* 1992, *47*, 267-277.
[68] Moon, D.-K.; Maruyama, T.; Osakada, K.; Yamamoto, T. Chemical Oxidation of Polyaniline by Radical Generating Reagents, O_2, H_2O_2–$FeCl_3$ Catalyst, and Dibenzoyl Peroxide. *Chem. Lett.* 1991, *20*, 1633-1636.
[69] Sun, Z.; Geng, Y.; Li, J.; Wang, X.; Jing, X.; Wand, F. Catalytic oxidization polymerization of aniline in an H_2O_2-Fe^{2+} system. *J. Appl. Polym. Sci.* 1999, *72*, 1077-1084.
[70] Toshima, N.; Yan, H.; Kajita, M.; Honda, Y.; Ohna, N. Novel Synthesis of Polyaniline Using Iron(III) Catalyst and Ozone. *Chem. Lett.* 2000, *29*, 1428-1429.
[71] Hua, X.; Zhang, Y. -Y.; Tang, K.; Zoua, G. -L. Hemoglobin-biocatalysts synthesis of a conducting molecular complex of polyaniline and sulfonated polystyrene. *Synth. Met*. 2005, *150*,1–7.
[72] Nabid, M. R.; Sedghi, R.; Jamaat, P. R.; Safari, N; Entezami, A. A. Synthesis of conducting water-soluble polyaniline with iron(III) porphyrin. *J. Appl. Polym. Sci.* 2006, 102, 2929–2934.
[73] Nabid, M. R.; Sedghi, R.; Jamaat, P. R.; Safari, N.; Entezami, A. A. Catalytic oxidative polymerization of aniline by using transition-metal tetrasulfonated phthalocyanine. *Appl. Catal. A: General* 2007, *328*, 52–57.
[74] Della Pina, C.; Rossi, M.; Ferretti, A. M.; Ponti, A.; Lo Faro, M.; Falletta, E. One-pot synthesis of polyaniline/Fe_3O_4 nanocomposites with magnetic and conductive behaviour. Catalytic effect of Fe_3O_4 nanoparticles *Synth. Met.* 2012, *162*, 2250-2258.

[75] Chiang, J. -C.; MacDiarmid, A. G. 'Polyaniline': Protonic acid doping of the emeraldine form to the metallic regime. *Synth. Met.* 1986, *13*, 193-205.

[76] Wei, Y.; Jang, G. W.; Hsueh, K. F.; Scherr, E. M.; MacDiarmid, A. G.; Epstein, A. J. Thermal transitions and mechanical properties of films of chemically prepared polyaniline. *Polymer*, 1992, *33*, 314-322.

[77] Pouget, J. P.; Jozefowicz, M. E.; Epstein, A. J.; Tang, X.; MacDiarmid, A. G. X-ray structure of polyaniline. *Macromolecules* 1991, *24*, 779-789.

[78] Focke, W. W.; Wnek, G. E.; Wei, Y. The influence of oxidation state, pH and counterion on the conductivity of polyaniline. *J. Phys. Chem.* 1987, *91*, 5813-5818.

[79] MacDiarmid, A. G.; Chiang, J.–C.; Richter, A. F.; Epstein, A. J. Polyaniline: a new concept in conducting polymers. *Synth. Met.* 1987, *18*, 285-290.

[80] Epstein, A. J.; Ginder, J. M.; Zuo, F.; Bigelow, R. W.; Woo, H.–S.; Tanner, D. B.; Richter, A. F.; Wang, W. –S.; MacDiarmid, A. G. Insulator-to-metal transition in polyaniline. *Synth. Met.*1987, *18*, 303-309.

[81] MacDiarmid, A. G. Polyaniline and polypyrrole: where are we headed?. *Synth. Met.* 1997, *84*, 27-34.

[82] Kitani, A.; Yano, J.; Kunai, A.; Sasaki, K. A conducting polymer derived from *para*-aminodiphenylamine. *J. Electroanal. Chem.* 1987, *221*, 69-82.

[83] Geniès, E. M.; Penneau, J. F.; Lapkowski, M.; Boyle, A. Electropolymerization Reaction Mechanism of *para*-aminodiphenylamine. *J. Electroanal. Chem.* 1989, *269*, 63-75.

[84] De Gennes, P. -G. Solutions de polymeres conducteurs. Lois d'echelles. *C. R. Acad. Sci.* 1986, *302*, 1-5.

[85] Oueiny, C.; Berlioz, S.; Perrin, F. –X. Carbon nanotube–polyaniline composites. *Prog. Polym.Sci.* 2014, *39*, 707-748.

[86] Shvartzman-Cohen, R.; Nativ-Roth, E.; Baskaran, E; Levi-Kalisman,Y; Szleifer, I; Yerushalmi-Rozen R. Selective dispersion of single-walled carbon nanotubes in the presence of polymers: the role of molecular and colloidal length scales. *J. Am. Chem. Soc.* 2004, *126*, 14850–14857.

[87] Breuer, O.; Sundararaj, U. Big returns from small fibers: a review of polymer/carbon nanotube composites. *Polym. Compos.* 2004, *25*, 630–645.

[88] Xie, X. L.; Mai, Y. W.; Zhou, X. P. Dispersion and alignment of carbon nanotubes in polymer matrix: a review. *Mater. Sci. Eng. R* 2005, *49*, 89–112.

[89] Huang, J. E.; Li, X. H.; Xu, J. C.; Li, H. L. Well-dispersed single-walled carbon nanotube/polyaniline composite films. *Carbon* 2003, *41*, 2731–2736.

[90] Baibarac, M.; Baltog, I.; Lefrant, S.; Mevellec, J. Y.; Chauvet, O. Polyaniline and carbon nanotubes based composites containing whole units and fragments of nanotubes. *Chem. Mater.* 2003, *15*, 4149–4156.

[91] Xu, J.; Yao, P.; Wang, Y.; He, F.; Wu, Y. Synthesis and characterization of HCl doped polyaniline grafted multi-walled carbon nanotubes core–shell nano-composite. *J. Mater. Sci.* 2009, 20, 517–527.

[92] Lu, X.; Dou, H.; Yang, S.; Hao, L.; Zhang, L.; Shen, L.; Zhang, F.; Zhang, X. Fabrication and electrochemical capacitance of hierarchical graphene/polyaniline/carbon nanotube ternary composite film. *Electrochim. Acta* 2011, *56*, 9224–9232.

[93] Sheng, Q.; Wang, M.; Zheng, J. A novel hydrogen peroxide biosensor based on enzymatically induced deposition of polyaniline on the functionalized graphene–carbon nanotube hybrid materials. *Sens. Actuators B* 2011, *160*, 1070–1077.

[94] Tseng, R. J.; Huang, J.; Ouyang, J.; Kaner, R. B.; Yang, Y. Polyaniline Nanofiber/Gold Nanoparticle Nonvolatile Memory. *Nano Lett.* 2005, *5*, 1077-1080.

[95] Xian, Y.; Hu, Y.; Liu, F.; Xian, Y.; Wang, H.; Jin, L. Glucose biosensor based on Au nanoparticles-conductive polyaniline nanocomposite. Biosens. *Bioelectron.* 2006, 21, 1996-2000.

[96] Sajanlal; P. R.; Sreeprasad, T. S.; Nair, A. S.; Pradeep, T. Wires, Plates, Flowers, Needles, and Core−Shells: Diverse Nanostructures of Gold Using Polyaniline Templates. *Langmuir* 2008, *24*, 4607-4614.

[97] Kang, E. T. Preparation of Nanosized Metallic Particles in Polyaniline. *J. Colloid Interface Sci.* 2001, *239*, 78-86.

[98] Mallick, K.; Witcomb, M. J.; Dinsmore, A.; Scurrell, M. S. Polymerization of Aniline by Auric Acid: Formation of Gold Decorated Polyaniline Nanoballs. *Macromol. Rapid Commun.* 2005, *26*, 232-235.

[99] Feng, X.; Mao, C.; Yang, G.; Hou, W.; Zhu, J. –J. Polyaniline/Au Composite Hollow Spheres: Synthesis, Characterization, and Application to the Detection of Dopamine. *Langmuir* 2006, *22*, 4384-4389.

[100] Feng, X.; Yang, G.; Xu, Q.; Hou, W.; Zhu, J. –J. Self-Assembly of Polyaniline/Au Composites: From Nanotubes to Nanofibers. *Macromol. Rapid. Commun.* 2006, *27*, 31-36.

[101] Kang, E. T.; Ting, Y. P.; Neoh, K. G.; Tan, K. L. Spontaneous and sustained gold reduction by polyaniline in acid solution. *Polymer,* 1993, 34, 4994-4996.

[102] Neoh, K. G.; Tan, K. K.; Goh, P. L.; Huang, S. W.; Kang, E. T. W.; Tan, K. L. Electroactive polymer–SiO_2 nanocomposites for metal uptake. *Polymer* 1999, *40*, 887-893.

[103] Wang, Y.; Liu, Z.; Han, B.; Sun, Z.; Huang, Y.; Yang, G. Facile Synthesis of Polyaniline Nanofibers Using Chloroaurate Acid as the Oxidant. *Langmuir* 2005, *21*, 833-836.

[104] Mallick, K.; Witcomb, M. J.; Scurrell, M. S. Polyaniline stabilized highly dispersed gold nanoparticle: an *in-situ* chemical synthesis route. *J. Mater. Sci.* 2006, *41*, 6189-6192.

[105] Pillalamarri, S. K.; Blum, F. D.; Bertino, M. F. Synthesis of polyaniline–gold nanocomposites using "grafting from" approach. *Chem. Comm.* 2005, 4584-4585.

[106] Busbee, B. D.; Obare, S. O.; Murphy, C. An improved synthesis of high-aspect-ratio gold nanorods. *J. Adv. Mater.* 2003, *15*, 414-416.

[107] Peng, Z.; Guo, L.; Zhang, Z.; Tesche, B.; Wilke, T.; Germann, D.; Hu, S.; Kleinermanns, K. Micelle-Assisted One-Pot Synthesis of Water-Soluble Polyaniline−Gold Composite Particles. *Langmuir* 2006, *22*, 10915-10918.

[108] Lahav, M.; Weiss, E. A.; Xu, Q.; Whitesides, G. M. Core−Shell and Segmented Polymer−Metal Composite Nanostructures. *Nano Lett.* 2006, *6*, 2166-2171.

[109] Xie, K; Li, J.; Lai, Y.Q,; Zhang, Z.; Liu, Y.; Zhang, G.; Huanget, H. Polyaniline nanowire array encapsulated in titania nanotubes as a superior electrode for supercapacitors. *Nanoscale* 2011, *3*, 2202–2207.

[110] Lin, Y. M.; Li, D. Z.; Hu, J.; Xiao, G.; Wang, J.; Li, W.; Fu, X. Highly efficient photocatalytic degradation of organic pollutants by PANI-modified TiO_2 composite *J. Phys. Chem. C* 2012, *116*, 5764–5772.

[111] Zhao, Z.; Zhou, Y.; Wan, W.; Wang, F.; Zhang, Q.; Lin Y. Nanoporous TiO_2 /polyaniline composite films with enhanced photoelectrochemical properties. *Mater. Lett.* 2014, *130*, 150-153.

[112] Fong, H.; Reneker, D. H. Electrospinning and formation of nanofibers. Salem DR editor, *Structure formation in polymeric fibers*, Munich: Hanser, 2001, 225–246.

[113] Hong, K. H.; Kang, T. J. Polyaniline–nylon 6 composite nanowires prepared by emulsion polymerization and electrospinning process. *J. Appl. Polym. Sci.* 2006, *99*, 1277-1286.

[114] Liu, Q.; Cui, Z.; Ma, Z.; Bian, S.; Song, W.; Wan, L. Morphology control of Fe_2O_3 nanocrystals and their application in catalysis. *Nanotechnology* 2007, *18*, 385605-385610.

[115] Willard, M. A.; Kurihara, L. K.; Carpenter, E. E.; Calvin, S.; Harris, V. G. Chemically prepared magnetic nanoparticles. *Int. Mater. Rev.* 2004, *49*,125-170.

[116] Jovalekic, C.; Zdujic, M.; Radakovic, A.; Mitric, M. Mechanochemical synthesis of $NiFe_2O_4$ ferrite. *Mater. Lett.* 1995, *24*, 365-368.

[117] Lee, S. –J.; Jeong, J. –R.; Shin, S. –C.; Kim, J. –C.; Kim, J. –D. Synthesis and characterization of superparamagnetic maghemite nanoparticles prepared by coprecipitation technique. *J. Magn. Magn. Mater.* 2004, *282*, 147-150.

[118] Rockenberger, J.; Scher, E. C.; Alivisatos, A. P. A New Nonhydrolytic Single-Precursor Approach to Surfactant-Capped Nanocrystals of Transition Metal Oxides. *J. Am. Chem. Soc.* 1999, *121*, 11595-11596.

[119] Lu, X.; Yu, Y.; Chen, L.; Mao, H.; Gao, H.; Wang, J.; Zhang, W.; Wei, Y. Aniline dimer–COOH assisted preparation of well-dispersed polyaniline–Fe_3O_4 nanoparticles. *Nanotechnology* 2005, *16*, 1660-1665.

[120] Kryszewski, M.; Jeszka, J. K. Nanostructured conducting polymer composites — superparamagnetic particles in conducting polymers. *Synth. Met.* 1998, *94*, 99–104.

[121] Wan, M. X.; Zhou, W.; Li, J. C. Composite of polyaniline containing iron oxides with nanometer size. *Synth. Met.* 1996, *78*, 27–31.

[122] Xiao, Q.; Tan, X.; Ji, L.; Xue, J. Preparation and characterization of polyaniline/nano-Fe_3O_4 composites via a novel Pickering emulsion route. *Synth. Met.* 2007, *157*, 784-791.

[123] Xue, W.; Fang, K.; Qiu, H.; Li, J.; Mao, W. Electrical and magnetic properties of the Fe_3O_4–polyaniline nanocomposite pellets containing DBSA-doped polyaniline and HCl-doped polyaniline with Fe_3O_4 nanoparticles. *Synth. Met.* 2006, *156*, 506–509.

[124] Aphesteguy, J. C.; Bercoff, P. G.; Jacobo, S. E. Preparation of magnetic and conductive Ni–Gd ferrite-polyaniline composite. *Physica B* 2007, 398, 200–203.

[125] Liu, G.; Freund, M. S. New Approach for the Controlled Cross-Linking of Polyaniline: Synthesis and Characterization. *Macromolecules*, 1997, *30*, 5660–5665.

[126] Andreatta, A.; Smith, P. Processing of conductive polyaniline-UHMW polyethylene blends from solutions in non-polar solvents. *Synth. Met.* 1993, *55*, 1017.

[127] Kang, Y.; Lee, M. –H.; Rhee, S. B. Electrochemical properties of polyaniline doped with poly(styrenesulfonic acid). *Synth. Met.* 1992, *52*, 319-328.

[128] Pud, A.; Ogurtsov, N.; Korzhenko, A.; Shapova, G. Some aspects of preparation methods and properties of polyaniline blends and composites with organic polymers. *Prog. Polym. Sci.* 2003, *28*, 1701–1753.

[129] Yang, S.; Ruckenstein, E. Processable conductive composites of polyaniline/poly(alkyl methacrylate) prepared via an emulsion method. *Synth. Met.* 1993, *59*, 1-12.

[130] Yoon, C. O.; Reghu, M.; Moses, D.; Heeger, A. J.; Cao, Y. Electrical transport in conductive blends of polyaniline in poly(methyl methacrylate). *Synth. Met.* 1994, *63*, 47-52.

[131] Juvin, P.; Hasik, M.; Fraysse, J.; Planès, J.; Pron, A.; Kulszewicz-Bajer, I. Conductive blends of polyaniline with plasticized poly(methyl methacrylate) *J. Appl. Polym. Sci.* 1999, *74*, 471–479.

[132] Hosoda, M.; Hino, T.; Kuramoto, N. Facile preparation of conductive paint made with polyaniline/dodecylbenzenesulfonic acid dispersion and poly(methyl methacrylate). *Polym. Int.* 2007, *56*, 1448–1455.

[133] Doshi, J.; Reneker, D. H. Electrospinning process and applications of electrospun fibers. *J. Electrost.* 1995, *35*, 151–160.

[134] Deitzel, J. M.; Kleinmeyer, J.; Hirvonen, J. K.; Beck Tan, N. C. Controlled deposition of electrospun poly(ethylene oxide) fibers. *Polymer* 2001, *42*, 8163–8170.

[135] Attout, A.; Yunus, S.; Bertrand, P. Electrospinning and alignment of polyaniline-based nanowires and nanotubes. *Polym. Eng. Sci.* 2008, *48*, 1661-1666.

[136] Zhang, Y.; Rutledge, G. C. Electrical Conductivity of Electrospun Polyaniline and Polyaniline-Blend Fibers and Mats. *Macromolecules* 2012, *45*, 4238–4246.

[137] Ondarcuhu, T.; Joachim, C. Drawing a single nanofibre over hundreds of microns. *Europhys. Lett.* 1998, *42*, 215–220.

[138] Feng, L.; Li, S.; Li, H.; Zhai, J.; Song, Y.; Jiang, L. A New Method for the Synthesis of Cycloheptenones by Rh^{I}-Catalyzed Intramolecular Hydroacylation of 4,6-Dienals. *Angew. Chem. Int. Ed.* 2002, *41*, 1221–1223.

[139] Ma, P. X.; Zhang, R. Synthetic nano-scale fibrous extracellular matrix. *J. Biomed. Mater. Research.* 1999, *46*, 60–72.

[140] Whitesides, G. M.; Grzybowski, B. Self-Assembly at All Scales. *Science* 2002, *295*, 2418–2421.

[141] Norris, I. D.; Shaker, M. M.; Ko, F. K.; MacDiarmid, A. G. Electrostatic fabrication of ultrafine conducting fibers: polyaniline/polyethylene oxide blends. *Synth. Met.* 2000, *114*, 109–114.

[142] MacDiarmid, A. G.; Jones, Jr. W. E.; Norris, I. D.; Gao, J.; Johnson, Jr. A. T.; Pinto, N. J.; Hone, J.; Han, B.; Ko, F. K.; Okuzaki, H. Electrostatically-generated nanofibers of electronic polymers. *Synth. Met.* 2001, *119*, 27-30.

[143] Lee, S. -H.; Yoon, J. -W. Continuous nanofibers manufactured by electrospinning technique. *Macromol. Research* 2002, *10*, 282-285.

[144] Son, W. K.; Youk, J. H.; Lee, T. S.; Park, W. H. The effects of solution properties and polyelectrolyte on electrospinning of ultrafine poly(ethylene oxide) fibers. *Polymer* 2004, *45*, 2959-2966.

[145] Frontera, P.; Busacca, C.; Antonucci, P.; Lo Faro, M.; Falletta, E.; Della Pina, C.; Rossi, M. Polyaniline nanofibers: towards pure electrospun PANI. *AIP Conf. Proc.* 2012, *1459*, 253.

[146] Jamshidian, M.; Tehrany, E. A.; Imran, M.; Jacquot, M; Desobry, S. Poly-Lactic Acid: Production, Applications, Nanocomposites, and Release Studies. *Comprehensive Rev. Food Sci. Food Safety* 2010, *9*, 552-571.

[147] Peng, S.; Zhu, P.; Wu, Y.; Mhaisalkar, S. G.; Ramakrishna, S. Electrospun conductive polyaniline–polylactic acid composite nanofibers as counter electrodes for rigid and flexible dye-sensitized solar cells. *RSC Adv.* 2012, *2*, 652–657.

[148] Picciani, P. H. S.; Medeiros, E. S.; Pan, Z.; Orts, W. J.; Mattoso, L. H. C.; Soares, B. G. Development of conducting polyaniline/poly(lactic acid) nanofibers by electrospinning. *J. Appl. Polym. Sci.* 2009, *112*, 744–753.

[149] Rahman, N. A.; Gizdavic-Nikolaidis, M.; Ray, S.; Easteal, A. J.; Travas-Sejdic, J. Functional electrospun nanofibres of poly(lactic acid) blends with polyaniline or poly(aniline-*co*-benzoic acid). *Synth. Met.* 2010, *160*, 2015-2022.

[150] Gizdavic-Nikolaidis, M.; Ray, S.; Bennett, J. R.; Easteal, A. J.; Cooney, R. P. Electrospun Functionalized Polyaniline Copolymer-Based Nanofibers with Potential Application in Tissue Engineering. *Macromol. Biosci.* 2010, *10*, 1424–1431.

[151] Kang, E.; Neoh, K.; Tan, T.; Khor, S.; Tan, K. Structural studies of poly(p-phenyleneamine) and its oxidation. *Macromolecules* 1990, *23*, 2918-2926.

[152] Xia, Y.; Wiesinger, J. M.; MacDiarmid, A. G. Camphorsulfonic Acid Fully Doped Polyaniline Emeraldine Salt: Conformations in Different Solvents Studied by an Ultraviolet/Visible/Near-Infrared Spectroscopic Method. *Chem. Mater.*1995, *7*, 443-445.

[153] Elsayed, A.; Eldin, M.; Elsyed, A.; Elazm, A.; Younes, E.; Motaweh, H. Synthesis and Properties of Polyaniline/ferrites Nanocomposites. *Int. J. Electrochem. Sci* 2011, *6*, 206.

[154] Snauwaert, P.; Lazzaroni, R.; Riga, J.; Verbist, J.; Gonbeau, D. Electronic properties of conjugated polymers III, Springer Series in Solid State Sciences, edited by Kuzmany, H.; Mehring, M.; Roth, S. Springer, Berlin, 1989, *91*, 301–304.

[155] Chung, D. D. L. Electromagnetic interference shielding effectiveness of carbon materials. *Carbon* 2001*, 39*, 279-285.

[156] Al-Saleh, M. H.; Sundararaj, U. Electromagnetic interference shielding mechanisms of CNT/polymer composites. *Carbon* 2009, *47*, 1738-1746.

[157] Wang, Y.; Jing, X. Intrinsically conducting polymers for electromagnetic interference shielding. *Polym. Adv. Technol.* 2005, *16*, 344–351.

[158] Epstein, A. J.; MacDiarmid, A. G. Polyanilines: From solitons to polymer metal, from chemical currosity to technology. *Synth. Met.* 1995, *69*, 179-182.

[159] Tantawy, H. R.; Aston, D. E.; Smith, J. R.; Young, J. L. Comparison of Electromagnetic Shielding with Polyaniline Nanopowders Produced in Solvent-Limited Conditions. *ACS Appl. Mater. Interfaces* 2013, *5*, 4648−4658.

[160] Zhang, Z.; Wan, M.; Wei, Y. Electromagnetic functionalized polyaniline nanostructures. *Nanotechnology* 2005, *16*, 2827-2832.

[161] Ziolo, R. F.; Giannelis, E. P.; Weinstein, B. A.; O'Horo, M. P.; Ganguly, B. N.; Mehrotra, V. M.; Russell, W.; Huffman, D. R. Matrix-Mediated Synthesis of Nanocrystalline γ-Fe2O3: A New Optically Transparent Magnetic Material. *Science* 1992, 257, 219-223.

[162] Gandhi, N.; Singh, K.; Ohlan, A.; Singh, D. P.; Dhawan, S. K. Thermal, dielectric and microwave absorption properties of polyaniline–$CoFe_2O_4$ nanocomposites. *Comp. Sci. Technol.* 2011, 71, 1754-1760.

[163] Hosseini, S. H.; Mohseni, S. H.; Asadnia, A.; Kerdari, H. Synthesis and microwave absorbing properties of

polyaniline/$MnFe_2O_4$ nanocomposite. *J. Alloys Compd.* 2011, *509*, 4682-4687.

[164] Belaabed, B.; Wojkiewicz, J. L.; Lamouri, S.; El Kamchi, N.; Lasri, T. Synthesis and characterization of hybrid conducting composites based on polyaniline/magnetite fillers with improved microwave absorption properties. *J. Alloys Compd.* 2012, *527*, 137-144.

[165] Saïdi, S.; Bouzitoun, M.; Mannaî, A.; Gmati, F.; Derouiche, H.; Belhadj A. M. Effect of PANI rate percentage on morphology, structure and charge transport mechanism in PANI–PVDF composites above percolation threshold. *J. Phys. D: Appl. Phys.* 2013, *46*, 355101-355111.

[166] Rahaman, M.; Chaki, T. K.; Khastgir, D. Polyaniline, ethylene vinyl acetate semi-conductive composites as pressure sensitive sensor. *J. Appl. Polym. Sci.* 2013, *128*, 161-168.

[167] Levin, Z. S.; Robert, C.; Feller, J. F.; Castro, M.; Grunlan, J. C. Flexible latex—polyaniline segregated network composite coating capable of measuring large strain on epoxy. *Smart Mater. Struct.* 2013, *22*, 015008-9p.

[168] Zhang, D. On the conductivity measurement of polyaniline pellets. *Polym. Test.* 2007, *26*, 9-13.

[169] C. Della Pina, M. Rossi, E. Falletta, Lecture notes in electrical engineering; 26 - In: Sensors and microsystems: proceedings of the 17th National Conference, Brescia, Italy, 5-7 February 2013, Springer, 2014 – p. 35-39.

[170] Della Pina, C.; Zappa, E; Busca, G.; Sironi, A.; Falletta, E. Electromechanical properties of polyanilines prepared by two different approaches and their applicability in force measurements. *Sens. Actuators B* 2014, *201*, 395-401.

INDEX

A

B

C

D

E

F

G

H

N

O

P

Q

R

S

T

U

V

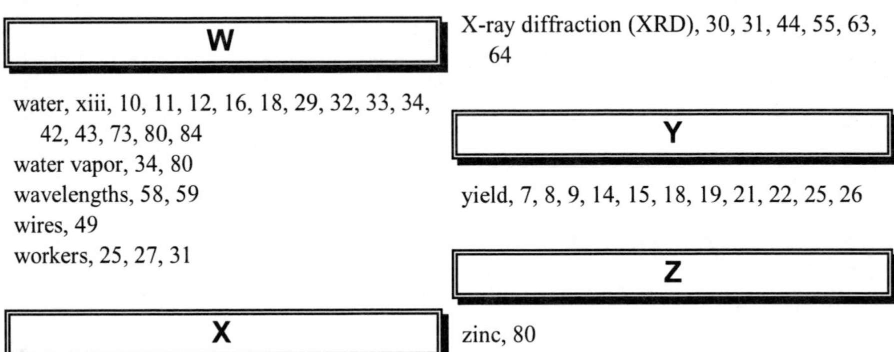

W

X

Y

Z